Hormones, Drugs & Social Behavior in Primates

Hormones, Drugs & Social Behavior in Primates

EDITED BY

HORST D. STEKLIS, Ph.D.
Department of Anthropology
Rutgers University
New Brunswick, New Jersey
and
Department of Psychiatry
Rutgers Medical School
University of Medicine and Dentistry of New Jersey
Piscataway, New Jersey

ARTHUR S. KLING, M.D.
Department of Psychiatry and Biobehavioral Sciences
School of Medicine
University of California, Los Angeles
Los Angeles, California
and
Psychiatry Service
Sepulveda Veterans Administration Medical Center
Sepulveda, California

SP MEDICAL & SCIENTIFIC BOOKS
a division of Spectrum Publications, Inc.
New York • London

Copyright © 1983 by Spectrum Publications, Inc.

All rights reserved. No part of this book may be reproduced in any form, by photostat, microform, retrieval system, or any other means without prior written permission of the copyright holder or his licensee.

SPECTRUM PUBLICATIONS, INC.
175-20 Wexford Terrace
Jamaica, NY 11432

Library of Congress Cataloging in Publication Data

Main entry under title:

Drugs, hormones, and social behavior in primates.

   Includes index.
   1. Hormones—Physiological effect. 2. Psychotropic drugs—Physiological effect. 3. Social behavior in animals. 4. Primates—Behavior. I. Steklis, Horst D. II. Kling, Arthur, 1929–   . [DNLM: 1. Social behavior—Congresses.
2. Behavior, Animal—Drug effects—Congresses.
3. Hormones—Physiology—Congresses. 4. Psychotropic drugs—Pharmacodynamics—Congresses. 5. Primates—Congresses. W3 IN5427L 8th 1980d / QL 775 D794 1980]
QP571.D78   1982   615'.78   82-6026
ISBN 0-89335-168-7

Printed in the United States of America

CONTRIBUTORS

T.G. Bidder, M.D.
Department of Psychiatry and
  Biobehavioral Sciences
School of Medicine
University of California, Los Angeles, and
Psychopharmacology Laboratory
Sepulveda Veterans Administration Hospital
Sepulveda, California

G.L. Brammer, Ph.D.
Department of Psychiatry
Sepulveda Veterans Administration Hospital
Sepulveda, California

G. Claus, M.D.
Department of Psychiatry
University of Medicine and Dentistry of New Jersey
Rutgers Medical School
Piscataway, New Jersey

C.L. Coe, Ph.D.
Department of Psychiatry and
  Behavioral Sciences
Stanford University School of Medicine
Stanford, California

J.A. Eberhart, Ph.D.
Department of Anatomy
Cambridge University
Cambridge, England

L. Engel, M.D.
Sepulveda Veterans Administration Hospital
Sepulveda, California

E. Geller, Ph.D.
Neurobiochemistry Laboratory
Brentwood Veterans Administration Hospital
Los Angeles, California

N. Harvey, Ph.D.
Department of Psychology
University of California, Riverside
Riverside, California

S.M. Howard, M.A.
Department of Psychology
Rutgers University
New Brunswick, New Jersey

C.K. Johnson, Ph.D.
Department of Psychiatry and
  Biobehavioral Sciences
School of Medicine
University of California, Los Angeles
Los Angeles, California

E.B. Keverne, Ph.D.
Department of Anatomy
Cambridge University
Cambridge, England

A.S. Kling, M.D.
Psychiatry Service
Sepulveda Veterans Administration Hospital
Sepulveda, California and
Department of Psychiatry and
  Biobehavioral Sciences
School of Medicine
University of California, Los Angeles
Los Angeles, California

G.W. Kraemer, Ph.D.
Primate Laboratory
University of Wisconsin
Madison, Wisconsin

S. Levine, Ph.D.
Department of Psychiatry and
  Behavioral Sciences
Stanford University
Stanford, California

G.S. Linn, M.A.
Department of Anthropology
Rutgers University
New Brunswick, New Jersey

M.T. McGuire, M.D.
Department of Psychiatry and
  Biobehavioral Sciences
School of Medicine
University of California, Los Angeles,
Los Angeles, California

W.T. McKinney, Jr. M.D.
Department of Psychiatry
University of Wisconsin Medical School
Madison, Wisconsin

R.E. Meller, Ph.D.
Department of Anatomy
Cambridge University
Cambridge, England

S.P. Mendoza, Ph.D.
Department of Sociology
Stanford University
Stanford, California

E.C. Moran, Ph.D.
Primate Laboratory
University of Wisconsin
Madison, Wisconsin

I. Munkvad, M.D.
Psychopharmacological Research Laboratory
St. Hans Mental Hospital, Department E
Roskilde, Denmark

H. Nieuwenhuijsen, Ph.D.
Department of Endocrinology, Growth, and
  Reproduction
Faculty of Medicine
Erasmus University
Rotterdam, The Netherlands

M.J. Raleigh, Ph.D.
Department of Psychiatry and
  Biobehavioral Sciences
School of Medicine
University of California, Los Angeles,
Los Angeles, California

A. Randrup, M.D.
Psychopharmacological Research Laboratory
St. Hans Mental Hospital, Department E
Roskilde, Denmark

D.E. Redmond, Jr. M.D.
Department of Psychiatry
Yale University School of Medicine
New Haven, Connecticut

E.N. Sassenrath, Ph.D.
Department of Psychiatry
School of Medicine
University of California, Davis
Davis, California

P.E. Schenck, Ph.D.
Department of Endocrinology, Growth and
  Reproduction
Faculty of Medicine
Erasmus University
Rotterdam, The Netherlands

A.K. Slob, Ph.D.
Department of Endocrinology, Growth, and
  Reproduction
Faculty of Medicine
Erasmus University
Rotterdam, The Netherlands

E.R. Smith, Ph.D.
Department of Physiology
Stanford University School of Medicine
Stanford, California

H.D. Steklis, Ph.D.
Department of Anthropology
Rutgers University, and
Department of Psychiatry
Rutgers Medical School
University of Medicine and Dentistry of New Jersey
New Brunswick, New Jersey

L. Tiger, Ph.D.
Department of Anthropology
Rutgers University
New Brunswick, New Jersey

A. Yuwiler, Ph.D.
Neurobiochemistry Laboratory
Brentwood Veterans Administration Hospital
Los Angeles, California

CONTENTS

INTRODUCTION.....................................1
    H. D. Steklis and A. S. Kling

Chapter 1
    VARYING INFLUENCE OF SOCIAL
    STATUS ON HORMONE LEVELS IN
    MALE SQUIRREL MONKEYS........................7
    C. L. Coe, E. R. Smith,
    S. P. Mendoza, and S. Levine

Chapter 2
    PLASMA TESTOSTERONE, SEXUAL
    AND AGGRESSIVE BEHAVIOR IN SOCIAL
    GROUPS OF TALAPOIN MONKEYS..................33
    E. B. Keverne, J. A. Eberhart,
    and R. E. Meller

Chapter 3
    STUDIES IN ADAPTABLILITY: EXPERIENTIAL,
    ENVIRONMENTAL, AND PHARMACOLOGICAL
    INFLUENCES..................................57
    E. N. Sassenrath

Chapter 4
    SOCIAL STATUS RELATED DIFFERENCES
    IN THE BEHAVIORAL EFFECTS OF
    DRUGS IN VERVET MONKEYS (CERCOPITHECUS
    AETHIOPS SABAEUS)...........................83
    M. J. Raleigh, G. L. Brammer,
    M. T. McGuire, A. Yuwiler,
    E. Geller, and C. K. Johnson

Chapter 5
    PROGESTERONE AND SOCIO-SEXUAL
    BEHAVIOR IN STUMPTAILED MACAQUES
    (MACACA ARCTOIDES): HORMONAL
    AND SOCIO-ENVIRONMENTAL INTERACTIONS........107
    H. D. Steklis, G. S. Linn,
    S. M. Howard, A. Kling, and
    L. Tiger

Addendum to Chapter 5
  A COMMENT ON CROSS-SPECIFIC
  COMPARISON OF THE EFFECTS OF
  PROGESTERONE TREATMENT ON SOCIAL
  BEHAVIOR..................................135
    L. Tiger

Chapter 6
  SOCIAL AND SEXUAL BEHAVIORS
  DURING THE MENSTRUAL CYCLE IN
  A COLONY OF STUMPTAIL MACAQUES
  (MACACA ARCTOIDES)........................141
    N. C. Harvey

Chapter 7
  EFFECTS OF CYPROTERONE ACETATE
  ON SOCIAL AND SEXUAL BEHAVIOR
  IN ADULT MALE LABORATORY HOUSED
  STUMPTAILED MACAQUES (MACACA
  ARCTOIDES)................................175
    A. K. Slob, P. E. Schenck,
    and H. Nieuwenhuijsen

Chapter 8
  EFFECTS OF METHAQUALONE ON SOCIAL-
  SEXUAL BEHAVIOR IN MACACA MULATTA.........205
    G. Claus and A. Kling

Chapter 9
  INFLUENCE OF AMPHETAMINE AND
  NEUROLEPTICS ON THE SOCIAL BEHAVIOR
  OF VERVET MONKEYS.........................237
    I. Munkvad and A. Randrup

Chapter 10
  EFFECTS OF DRUGS ON THE RESPONSE
  TO SOCIAL SEPARATION IN RHESUS
  MONKEYS...................................249
    W. T. McKinney, Jr.,
    E. C. Moran, and G. W. Kraemer

Chapter 11
   SOCIAL EFFECTS OF ALTERATIONS
   IN BRAIN NORADRENERGIC FUNCTION
   ON UNTREATED GROUP MEMBERS....................271
     D. E. Redmond, Jr.

Chapter 12
   STRATEGIC PSYCHOPHARMACOTHERAPY:
   THE THERAPEUTIC USE OF MEDICATION
   IN FAMILY SYSTEMS.............................281
     L. Engel and T. G. Bidder

Chapter 13
   A MODEL FOR STUDYING DRUG USE,
   AND EFFECTS IN DYADIC INTERACTIONS............321
     M. T. McGuire

Index..353

ACKNOWLEDGMENTS

The organizers and participants of the 1980 Pisa Symposium are indebted to Dr. Silvana Borgognini of the University of Pisa, who served as local guest organizer and aided us in the organization of the symposium as a satellite to the International Primatological Society Congress in Florence. On this side of the Atlantic, we are immeasurably grateful to Ms. Sylva Grossman for a scrutinizing editorial eye, for invaluable assistance in the arduous task of collecting the papers into a coherent volume, and for preparing the subject index. We are especially indebted to Ms. Tovah Hollander and Ms. Laurie Phillip for countless hours spent editing, formatting, and typing the manuscripts.

DEDICATION

This volume is dedicated to Dr. Fred Elmadjian, Retired Deputy Director for Manpower and Training, National Institute of Mental Health. His energy and devotion to the founding and development of interdisciplinary research training in mental health has earned him the enduring gratitude of generations of behavioral scientists.

Hormones, Drugs & Social Behavior in Primates

INTRODUCTION

Despite the vast amount of information that exists on the effects of psychotropic drugs on human behavior and psychopathology, there is little reported on how these agents manifest their influence within a social context. Much of what we know about basic behavioral mechanisms has been obtained from studies which have focused on motor, cognitive, or perceptual functions. In humans, the focus has been on the study of the individual who is taking the substance, with particular emphasis on affect, perception, cognition, or the modification of psychopathology.

When we (the editors) received notice that the International Congress of Primatology was to meet in July 1980 in Florence, Italy, we suggested to the organizers a symposium on the "effects of drugs and hormones on social behavior in non-human primates." We were delighted at the positive responses received from many of our colleagues, who participated despite the absence of support for travel and per diem expenses.

To the editors, the topic of the symposium was a natural outgrowth of a series of seminars and informal discussions over several years among faculty and trainees of the NIMH sponsored interdisciplinary research training program at Rutgers University and Medical School, focusing on the neurobiology of primate social behavior. While some of the faculty and trainees were involved in studying the influence of social and environmental factors on the behavior of brain-lesioned subjects, others were examining the influence of the physical and social environment on pharmacologically induced behavior changes and hormone-environment interactions.

Although most of the contributions to this volume were presented at the Pisa symposium, a number of additional papers were solicited to broaden the scope of the volume, especially in human socio-psychopharmacology and related methodological issues in this complex area of research.

It is interesting to note that despite the intense interest and prodigious number of studies on psychoactive drugs in animals and man, there has been relatively little interest in how drug-induced changes in treated subjects are influenced by the social-environmental setting and, conversely, how these treated individuals affect others around them. Since it has been well established that many social and environmental variables have profound effects on endocrine and metabolic functions, it is not surprising that such changes should affect the action of certain drugs and hormones in the central nervous system and consequently the behavior of the subject. The most common examples of this interaction are found in the difference between the actions of many psychoactive drugs in normal vs. psychopathological conditions. This issue has been addressed in the paper by Sassenrath, in a study of the interaction between early experience, social stress, and the long-term effects of marijuana, and by McKinney and Kraemer's studies of early social deprivation and the effects of anti-depressants on behavioral pathology.

In the process of treating chronic schizophrenic patients with neuroleptics, are we facilitating the tendency for social withdrawal in some patients? Some suggestions that this may be the case were presented by Munkvad and Randrup. This possibility has profound implications for the long-term care of chronic schizophrenia patients with respect to their rehabilitation and re-socialization potential. Further, how does a social group react to the presence of asocial individuals? If our present studies of socially withdrawn monkeys, following lesions of either amygdala, temporal pole or posterior orbital cortex, are relevant (Kling and Steklis, 1976) we might anticipate agression toward such individuals and attempts by the normals to exclude them from their territory. Perhaps the hostility exhibited by residents of urban communities to chronic schizophrenic patients resettling in their midst is an example in man of this effect.

The marked differences observed between the behavioral effects of a drug on an individually-caged subject and the same subject in a social setting are discussed by Claus and Kling in an examination of the influence of a commonly used street drug (Methaqualone) on social-sexual behavior in rhesus monkeys (Macaca mulatta). Reference is also made to how differences in the composition of the group affect the behavior of the treated subject. Of particular interest in this study was the increased sexual behavior observed in Methaqualone-treated males, which may substantiate the subjective reports from street users that this drug is an aphrodisiac.

The chapter by Steklis and colleagues demonstrates that discrepancies between studies demonstrating differing effects of the same drug treatment may be based on environmental differences. They studied the copulatory behavior of male macaques (M. arctoides) toward females treated with a long-acting progestational agent, medroxyprogesterone acetate (Depo-provera), when males had limited access to females, as in laboratory behavior tests, and when subjects were part of a stable social group in a semi-free-ranging setting. Their results point to the dangers of drawing broad conclusions from observations in one socio-environmental setting.

While it has been shown that environmental factors may influence neuroendocrine processes, particularly the ACTH-adrenal axis, much less is known about the LH-gonadal system, especially with respect to the regulation of social-sexual behavior. Studies of primate behavior patterns in laboratory and field conditions have established that breeding systems are under a variety of control mechanisms, which range from variations in endogenous hormone concentrations and seasonal variations in food supply, light, and rainfall, to social mechanisms like dominance and kinship. Several papers in this volume address such mechanisms. Slob and co-workers, for example, address the relative importance of social and hormonal variables in the maintenance of sexual behavior following administration of an anti-androgen (Cyproterone acetate) to male stumptail macaques (M. arctoides). In a review of the influence of dominance relations on adrenal and gonadal hormones in male squirrel monkeys (Saimiri sciureus), Coe et al. discuss how this influence is modified by seasonality and different social conditions. Keverne's experiments with talapoins (Miopithecus talapoin) on the interactions between social rank, aggressive behaviors, and hormone levels demonstrate, for example, how social factors can inhibit LH release in estrogenized females. The study of stumptail macaques (M. arctoides) by Harvey focuses on the interaction between kinship, rank, and stage of menstrual cycle in the regulation of sexual interactions in a group.

Despite the wide variations that exist among primate species, including man, in patterns of affiliative bonds, it is widely accepted that these bonding systems are crucial to survival. One species, e.g., the gibbon, lives in small family groups with only a mated pair and offspring, while other species such as baboons may group into large heterosexual bands or harems controlled by an adult male. Clearly, natural selection must have favored individuals with mechanisms that motivated and enabled them to maximize their fitness through the formation of social bonds.

Kling and Steklis along with colleagues and students have suggested on the basis of their experiments and those from other laboratories that three brain areas, i.e., the amygdaloid nucleus, temporal pole, and posterior orbital frontal cortex, are necessary for the maintenance of affiliative bonds in primates (Kling and Steklis, 1976). When any one of these structures is bilaterally ablated, the subject, depending on the experimental setting, will show a marked decrease in affiliative behaviors, as in a confined group, or total social isolation and eventual death in free-ranging environments.

Recent studies by Raleigh and colleagues have shown that plasma levels of free tryptophan may be reflective of the degree of affiliative behaviors in the vervet (Cercopithecus aethiops). Further, these studies, summarized in this volume, indicate that drugs which enhance central serotonergic activity resulted in increases in affiliative types of behaviors, while drugs that reduce serotonergic activity reduced these behaviors. It is interesting that the brain structures previously identified as being necessary for the maintenance of affiliative behavior also receive strong serotonergic projections from cells of origin in the nucleus raphe. This raises the possibility of a linkage between plasma tryptophan, central serotonin, and the localized brain areas mentioned as a neuro-chemical system for the regulation of affiliative behavior.

Redmond et al. have focused on the influence of central noradrenergic measures on the behavior of both the treated and untreated members of social groups of stumptail macaques. Previous work by this group has suggested a link between peripheral measures of MAO activity and intensity of social bonding among individuals in free-ranging macaques (M. mulatta). Thus, it appears that we are beginning to gain some insight into the complex physiological and biobehavioral substrates for the observed bonding systems in various species of primates.

The complex interactions between the influence of psycho-active drug treatments or hormonal manipulations and socio-environmental factors on the neural regulation of behavior are only beginning to be explored, and there is a great need for heuristic models. The chapters by Engel and Bidder and by McGuire are particularly courageous attempts at such model building.

The editors hope that the experimental and theoretical papers in this volume will provide impetus and challenge for future investigation and lead to eventual understanding of the biological mechanisms subserving those adaptive qualities of human behavior that are the product of millions of years of evolution.

Horst D. Steklis
and
Arthur S. Kling

REFERENCE

Kling, A. and Steklis, H.D. A neural substrate for affiliative behavior in nonhuman primates. Brain, Behavior and Evolution 13, 216-238 (1976).

Chapter 1
Varying Influence of Social Status on Hormone Levels in Male Squirrel Monkeys

*C.L. Coe, E.R. Smith,
S.P. Mendoza, and S. Levine*

INTRODUCTION

The complexity of dominance relations in primate social groups has been the subject of much discussion over the last two decades (Hall, 1965; Gartlan, 1968; Bernstein, 1970; van Kreveld, 1970; Rowell, 1974; Deag, 1977). Most authors conclude that the concept of dominance is necessary and meaningful for understanding primate social behavior although they emphasize that dominance status must be considered as a multi-determined attribute which can vary across time and different environmental conditions. Given this variability, however, one of the most difficult questions to answer has been how dominance behavior is related to internal physiological functions. Evidence of hormonal involvement in the expression of aggression and territoriality in other species has led investigators to evaluate the association between dominance and hormone levels in primates. While it appears that high hormone output is not necessary for the expression of dominance behavior in primates (Epple, 1978; Green et al., 1972; Bernstein et al., 1979a), a number of studies have demonstrated that dominance status can influence circulating levels of adrenal and gonadal hormones in several primate species (Candland and Leshner, 1974; Rose et al., 1974; Coe et al., 1979; Dixson, 1980). The latter findings have had a prevailing influence on discussions of dominance despite periodic reports of failures to find an effect of dominance on hormone levels (Eaton and Resko, 1974).

It is apparent that further research needs to be conducted in this area in order to reconcile the discrepant data and to develop a more complete understanding of the relationship between social behavior and endocrine physiology. Some of the differences between the previous hormonal studies may, in fact, be parsimoniously explained on the basis of variables which are known to influence the manifestation of dominance. With respect to gonadal hormones in males, for example, dominance may have a greater association with hormonal output during the mating season; thus, the failure of Eaton and Resko (1974) to find a correlation between testosterone levels and social rank in Japanese macaques may have been due to the collection of blood samples during the nonmating season. The effect of dominance on hormone levels may also be accentuated during competitive situations and following agonistic encounters, such as during the formation of new social groups (Bernstein et al., 1979b); therefore, the association between hormones and behavior may be less overt in stable groups after social roles have been established (Green et al., 1972; Bernstein et al., 1979a; Gordon et al., 1979). Some of the discrepancies concerning the influence of dominance on the pituitary-adrenal system can also be partially resolved. Several studies have reported higher adrenal output in subordinates (Sassenrath, 1970; Chamove and Bowman, 1978), which concurs with research on rodents, while other studies have indicated that dominant animals can show higher levels of corticosteroids under certain circumstances (Candland and Leshner, 1974; Coe et al., 1979). It appears that we must distinguish situations of social stress and overcrowding, which accentuate subordination, from other situations such as the exposure to novel environments or external threats, where dominant animals may be more vigilant and reactive resulting in greater adrenal activation.

The lack of resolution on these issues and the need to determine whether the proposed explanatory principles are truly efficacious led us to conduct a series of studies examining the relationship between dominance and hormone levels in male squirrel monkeys. We were interested in evaluating the effect of social relations on plasma cortisol and testosterone levels and in determining the most salient variables which influence endocrine function in this species. Adrenal and gonadal hormones were assessed simultaneously in each experiment, and the studies focused on the influence of factors, such as subject familiarity, group size, and biorhythms, which are known to affect the manifestation of dominance. As will be shown, these studies indicated that it is imperative to take a number of psychological variables into account when conducting endocrine studies on a social organism.

Several unique features of the biology of squirrel monkeys should be mentioned prior to discussing the results of these studies, however. The most important consideration for the study of endocrine function in this species is that both males and females secrete extremely high levels of steroid hormones (Brown et al., 1970; Wolf et al., 1977; Coe et al., 1978; Mendoza et al., 1978c; Wilson et al., 1978). It is premature to speculate on the functional significance of these high hormone levels, but suffice it to say for the purposes of the current discussion that high steroid hormone levels are a taxonomic characteristic of most small New World primates (Abbott and Hearn, 1978; Dixson et al., 1980; Nagle et al., 1980). In addition, the high hormone levels are not due to correspondingly high levels of binding proteins; rather they may be a response to target cell receptors that are comparatively insensitive to hormones (Brown et al., 1970; Chrousos et al., 1981).

Another biological feature of importance is that squirrel monkeys are seasonal breeders. Annual variation in hormones and behavior occurs under both laboratory and natural conditions, and mating typically occurs for three to four months of the year (Rosenblum, 1968; Baldwin, 1970; Coe and Rosenblum, 1978; Mendoza et al., 1978c). Peak hormone levels usually occur during the mating period in both males and females, and there may be an intensification of male dominance interactions at this time (Coe and Levine, 1981). Finally, it should be mentioned that squirrel monkeys associate in large, multi-male troops under natural conditions, and their gregarious nature permits the comparatively easy formation of new social groups in the laboratory. Linear dominance hierarchies are established fairly quickly, and aggression subsides to be replaced by more ritualized dominance interactions, which include a species-specific genital display and spatial displacement.(1)

DOMINANCE RELATIONS IN PAIR-HOUSED MALES

Our initial interest in exploring the influence of dominance on endocrine activity began during an experiment on adrenal and gonadal responses to stress when it became apparent that dominance relations between the subjects were affecting basal hormone levels (Coe et al., 1979). We had housed adult males in pairs for pragmatic reasons and quickly discovered that the dominant male in each cage was showing higher plasma levels of cortisol and testosterone than his partner.

This experiment was conducted on 16 wild-caught adult males of Bolivian origin. The males were in the heavier part of their annual weight variation (mean weight = 1270 g), and the blood samples were collected during the mating season of our laboratory which typically occurs from December to March. Each male was paired with a previously unfamiliar partner and housed in a small wire mesh cage (.61 x .46 x .91 m). The pairs were allowed to acclimate to these housing conditions for three weeks, and then blood samples were collected at weekly intervals from each male for five weeks. Blood sampling began at 0800, and samples were collected consecutively from each subject out of view of the remaining monkeys. To minimize the potential effect of immediate disturbance, samples were obtained in less than 2.0 - 3.5 min per animal through the use of ether anesthesia and the rapid collection of 1.0 ml samples via cardiac puncture. In addition, blood samples were collected from only four of the eight pairs on a given sampling day to reduce the total time needed for blood collection. Thirty minutes after each monkey was sampled, a second blood sample was collected to determine the hormonal response to the stress of the prior sampling procedure.(2)

Following the weeks of blood-sampling, dominance relations were assessed in each pair utilizing a series of behavioral tests designed to evaluate the 'priority of access to desired but limited incentives' (Figure 1). Five tests were employed because measures of dominance can be equivocal if motivational tendencies vary across subjects. As can be seen in Figure 1, the tests included (1) two different water competition tests conducted on separate days after six hours of water deprivation; (2) a food competition test conducted after six hours of food deprivation; (3) an assessment of sociosexual behavior when a female was housed near the males' cage; (4) an evaluation of dominance interactions related to the utilization of vertical height. Since squirrel monkeys are arboreal monkeys, they often use height to symbolically convey dominance, especially during aggressive interactions (Bailey, personal communication; Clark and Nakashima, 1972). The occurrence of genital displays (Ploog et al., 1963) during the tests was used to verify the designation of rank based on a composite score from the five test measures.

Using this approach it was possible to determine the dominant and subordinate member in every pair, and the mean levels of plasma cortisol and testosterone for the dominant and subordinate males are shown in Figure 2. In all eight pairs, dominant males had higher steroid hormone levels than did their subordinate partners. Basal testosterone values averaged 180 ng/ml in dominant males, significantly above the average 85 ng/ml in subordinate males, $F(1,12) = 12.34$, $p < .01$. Although testosterone values varied across pairs, a typical characteristic of dominant males was the more frequent occurrence of high hormone surges up to 473 ng/ml across the five-week sampling period. Similarly, despite high basal levels of adrenal corticosteroids in all males, the average cortisol level of 228 µg/100 ml in dominant males was significantly above the average 174 µg/100 ml in subordinate males, $F(1,12) = 14.11$, $p < .01$.

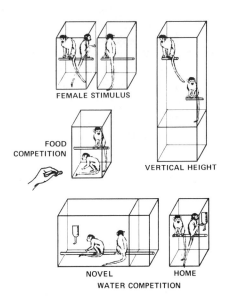

Figure 1. Five behavioral tests used to assess dominance relations in pairs of male squirrel monkeys. Cages are drawn on a smaller scale than actual size.

The hormonal responses 30 min after basal sampling were also influenced by dominance status, or more specifically by the different level of circulating hormones in dominant and subordinate males at the time of stress exposure. The increment in plasma cortisol at 30 min averaged 41 µg/100 ml in dominant males as compared to 76 µg/100 ml in subordinate males, $t(14) = 9.16$, $p < .001$, reflecting a negative correlation between high basal levels and the magnitude of the stress response ($r = -.40$). Similarly, dominant and subordinate males showed a different type of testosterone response following stress. Squirrel monkeys usually show a transient elevation in testosterone levels after stress, as do other organisms (e.g., Nieschlag and Wickings, 1980), before undergoing the more typically reported stress-induced suppression in hormone levels.

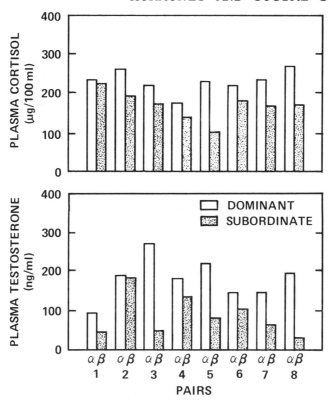

Figure 2. Plasma cortisol and testosterone levels in eight pairs of squirrel monkeys. Each bar represents the mean of five weekly samples.

However, at 30 min post-stress the subordinate males showed an average elevation of 28 ng/ml, whereas the dominant males showed an average elevation of only 5 ng/ml. This difference later proved to be due to a shorter duration of the testosterone elevation in dominant males, which occurs when basal levels prior to stress are relatively high.

EFFECT OF SEASONALITY

In the next experiment we examined the influence of dominance status at different times of the year and also extended our evaluation of the different stress responses in dominant and subordinate males. Four pairs of males were tested in the mating season (February-March), and three different pairs were tested in the nonmating season

(August-October). A paradigm similar to that of the previous experiment was utilized. Unfamiliar males were housed together as pairs, and after three weeks of acclimitization, weekly blood samples were collected at 0800 in order to determine basal and stress hormone levels. During both seasons, stress samples were collected at three time points -- 15, 30 and 45 min after basal sampling. In addition, during the nonmating season second samples were also collected at 240 min after basal sampling. Across the weeks of the study, each subject was sampled twice at all time points in a balanced order design. Thus, the data on basal levels portrayed in Figure 3 are derived from six weekly samples per male in the mating season and eight weekly samples per males in the nonmating season. Blood samples were also collected from two castrated males in the nonmating season to determine the relative contribution of adrenal androgens to the stress-induced increase of testosterone from the testes.

As can be seen in Figure 3, the basal levels of cortisol and testosterone in the mating season basically replicated the previous experiment. Dominant males tended to have higher hormone levels than did subordinate males, although variable hormone levels in one of the four pairs prevented the attainment of statistical significance. But more striking than the difference between dominant and subordinate males in the mating season was the marked drop in hormone levels during the nonmating season. Basal testosterone values decreased from an average 172 ng/ml in dominant males and 112 ng/ml in subordinate males during the mating season to an average 33 ng/ml and 18 ng/ml in dominant and subordinate males, respectively, during the nonmating season. At these lower hormone levels during the nonmating season, the likelihood of dominant males to show higher testosterone values was clearly reduced. Yet it is noteworthy that dominant males did continue to have higher testosterone levels on 17 of the 24 basal samples collected from the three pairs at this time of year.

Adrenal hormone secretion was also lower during the nonmating season which concurs with reports on other seasonal breeders (Christian,

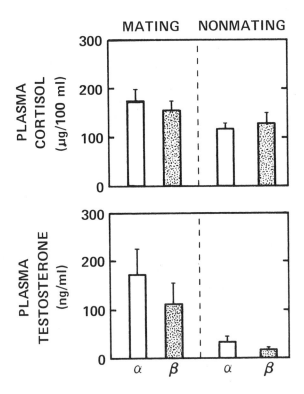

Figure 3. Mean (+SE) hormone levels for dominant and subordinate males at different times of the year. Mating season values are based on six weekly blood samples from four pairs; nonmating season values are based on eight weekly blood samples from three pairs.

1962; Parkes and Deanesley, 1966). When basal levels of plasma cortisol were higher during the mating season, dominant males tended to have higher values than subordinates (on 19 of the 24 basal samples collected from the four pairs). Following the decline in cortisol levels during the nonmating season, however, the association between dominance and hormones was not apparent.

The testosterone response to stress also showed a marked shift across the year (Figure 4). During the mating season both dominant and subordinate males showed a transient elevation in plasma testosterone following the stress of basal sampling, as observed in the previous experiment.

16   COE ET AL.

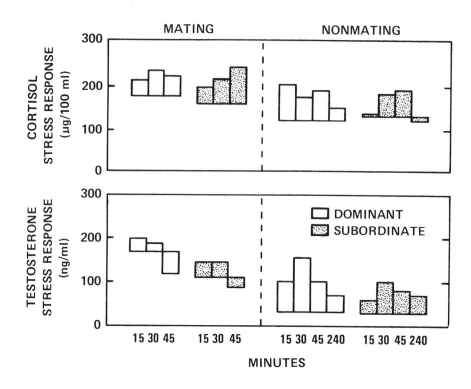

Figure 4. Mean changes in plasma cortisol and testosterone at different time points following basal blood sampling. The hormonal response is portrayed with respect to average basal levels of dominant and subordinate males in the two seasons.

By 45 min post-stress the testosterone levels had usually declined below basal values, especially if the preceding circulating levels had been high. In contrast, during the nonmating season the elevation in testosterone titers continued to persist for as long as four hours after basal sampling. Data collected on the castrated males indicated that this seasonal difference in the lability of the testosterone response reflected changes of gonadal rather that adrenal origin. Androgen levels in castrated males averaged only .15 ng/ml, and the stress increment in castrated males, presumably of adrenal origin, was negligible (.06 ng/ml). Unlike the testosterone response to stress, changes in the adrenocortical response were not apparent across the year (Figure 4); during the first 45 min after basal sampling, plasma cortisol levels typically rose an average of 53 µg/100 ml, and by four hours cortisol levels were returning to basal levels.

GROUP FORMATION AND FEMALE STIMULI

These data on pair-housed males led us to evaluate the possible influence of dominance on hormone levels in three male groups and also to examine the males' hormonal response to the introduction of female monkeys (Mendoza et al., 1979). Previous studies on other primates, including macaques and talapoin monkeys, as well as on other mammals, have indicated that males typically show an increase in testosterone levels in the presence of females, although the response may not occur in subordinate males (Bernstein et al., 1977; Liptrap and Raeside, 1978; Keverne, 1979). Our study was initiated in June during the nonmating season because we were also interested in assessing the hormonal responses to the induced-breeding which often occurs following group formations, even in the nonmating time of year (Coe and Levine, 1981). In the first phase of the study, nine male squirrel monkeys of Bolivian origin were initially housed alone for four weeks. The males were then merged into three groups, each consisting of three previously unfamiliar males, and housed in identical wire-mesh cages (1.8 X 1.8 X 1.2 m). One month after the formation of the male groups, five females of Bolivian origin were added to each cage.

Two blood samples were collected from each male during the Alone phase, and then blood samples were collected from the males at weekly intervals during the All-Male and Male-Female phases (Figure 5). All samples were collected at 0800, and the first samples in the social phases were obtained at 24 h after group formations. Male dominance relations were determined by observing each subject five min per day, three days per week across the study. Dominance rank was designated by generating social matrices of the genital display and manual grabbing hierarchies. The statistical analyses have been

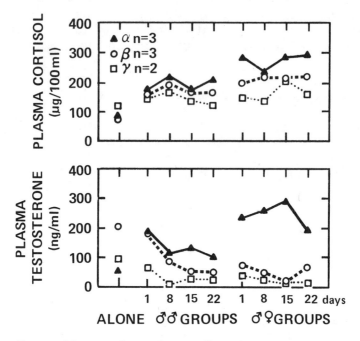

Figure 5. Mean levels of plasma cortisol and testosterone for males of each dominance rank during three phases of group formation.

based on a comparison of the hormonal responses of only the alpha and beta males, because the most subordinate male of one group had to be removed due to illness during the All-Male phase. Data for the two gamma males are shown graphically.

As usually occurs in small captive groups of squirrel monkeys, a linear dominance hierarchy developed rapidly in each group of males and the hierarchy remained constant across the study. Dominant males directed more genital displays and more manual grabbing at subordinates, and alpha males were the recipients of these behaviors only infrequently. Hormone levels in the Alone phase were not predictive of subsequent rank and thus the hormone-behavior relationship which emerged was due primarily to the attainment of a certain dominance status. Position in the dominance hierarchy had a significant effect on both cortisol and testosterone levels (Figure 5). Analysis of the mean testosterone values across the three phases revealed a significant interaction between dominance status and conditions, $F(2,8) = 17.06$, $p<.01$; alpha males showed an increase during each social phase while beta and gamma males underwent decreases with each successive phase. The increases in the plasma testosterone levels of alpha males, which began during the All-Male phase, were particularly striking during the Male-Female phase because hormone levels rose dramatically within 24 h after the introduction of females.

Unlike the selective testosterone response, adrenal output appeared to reflect an influence of both dominance and general agitation. Following group formation, cortisol levels increased significantly in all males and were highest while females were present, $F(2,8) = 57.32$, $p<.01$. The higher cortisol levels in the presence of the females may have been due in part to the onset on mating activity which was induced by the group formations even though the study was conducted in the nonmating season. In spite of this nonspecific effect of heightened arousal on adrenal secretion, dominant males secreted the highest level of plasma cortisol, as would be expected from the studies on pair-housed males, although the effect was not significant until after the introduction of females.

INFLUENCE OF GROUP SIZE

Based on the previous studies, we knew that dominance relations influenced hormone levels in both pairs and triads and that dominance rank could differentially affect the hormonal response to introducing females, but it remained to be determined whether these effects would persist in larger groups. Two additional groups have now been evaluated, one with four males and the other with six males of the Bolivian variety. The groups were formed just prior to the laboratory mating season and observed for three months from November through early February. Each male was observed for 30 min

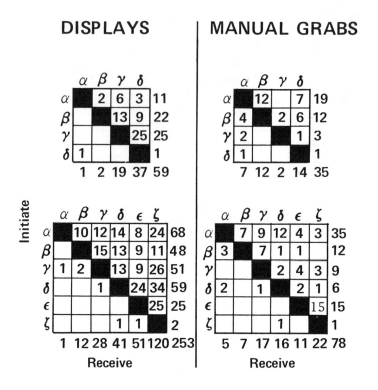

Figure 6. Dominance hierarchies in four-male and six-male groups. Data are based on six hours of observation per male across a three-month period.

per week for a total of six hours across the study. Blood samples (1 ml) were collected at biweekly intervals across this period. Six samples were collected from each monkey in the four-male group and seven samples were collected from each monkey in the six-male group at 0800.

In both groups the males established linear dominance hierarchies immediately after being housed together and the initial rank orders did not change across the study. As can be seen in Figure 6, higher-ranking males directed more genital displays at subordinate males and only rarely received displays from low-ranking animals. Observations of manual grabbing, a low level aggressive behavior related to spatial displacement, confirmed this assessment of the dominance hierarchies, in keeping with previous studies which have emphasized

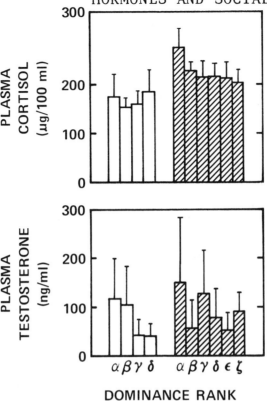

Figure 7. Mean (+SD) levels of plasma cortisol and testosterone for males of each dominance rank in four-male and six-male groups.

the significance of genital displays as a measure of dominance in squirrel monkeys. However, because middle-ranking animals both received and directed dominance behavior, social relations in these groups continued to be fairly complex, which may partially account for the outcome of the hormonal evaluations.

Despite the occurrence of linear dominance relations in both groups, the relationship between rank and hormone levels was not as clearly delineated as in pairs and triads (Figure 7). In neither group was there a simple correlation between a specific dominance rank and testosterone levels. In the four-male group alpha and beta males had significantly higher levels than did gamma and delta males, $F(3,15) = 5.24$, $p < .01$, but this was due primarily to the onset of seasonal testosterone surges which were not shown by the two low-ranking males. In the six-male group, testosterone levels did not differ significantly between males, although the alpha male again showed the highest seasonal surges in testosterone.

Plasma cortisol levels were also not systematically affected by dominance rank in either group. In the four-male group, plasma cortisol levels did not differ significantly between subjects; and in the six-male group, only the alpha male showed a significantly higher level of plasma cortisol, $F(5,30) = 6.74$, $p < .01$. It was of interest, however, that every monkey in the six-male group had higher levels of plasma cortisol than did monkeys in the four-male group. While it was not possible to determine the exact cause for this difference, it is likely that the more frequent dominance interactions in the larger group partially accounted for this higher overall level of adrenal activation. During the course of the three-month observation, only 59 genital displays were observed in the four-male group in contrast to a total of 253 displays in the six-male group.

GENERAL DISCUSSION

In overview, we believe that these studies provide convincing evidence that dominance relations can exert a strong influence on the endocrine physiology of squirrel monkeys under certain specifiable conditions. The effect of social status is most pronounced during the mating season when dominant males show significantly higher plasma levels of testosterone. Differences in hormone levels are also most clearly shown in pairs of males, or in triads exposed to females. The occurrence of higher testosterone titers in dominant males concurs with previous studies on other primate species (Rose et al., 1974; Dixson, 1980), and since the differences in hormone levels were sustained over considerable periods of time, the association is not due simply to initial agonistic encounters. At this point, it is not possible to specify whether there are cumulative benefits accrued by maintaining higher testosterone output, but it has already been shown that higher testosterone levels can facilitate ejaculatory reponses in this species (Chen et al., 1981). Higher testosterone titers may also enhance certain behavioral predispositions, as evidenced by the involvement of testosterone in the expression of some scent-marking behaviors in the squirrel monkey (Hennessy et al., 1980). In addition, as shown in the current work, changes in circulating levels of testosterone can influence the hormonal responses to stress.

Higher-ranking males in pairs and triads of squirrel monkeys also showed greater adrenocortical secretion. This finding is not in keeping with a number of previous studies on primates (Sassenrath, 1970; Manogue et al., 1975; Chamove and Bowman, 1978), but as mentioned in the Introduction, the effect of dominance on pituitary-adrenal activity appears to vary depending upon social and housing conditions. In particular, one must be cognizant of whether the animals are being observed in overcrowded situations. In our studies plasma cortisol levels proved to be sensitive to several factors in addition to dominance, as exemplified by higher levels in all males (1) during the mating season, (2) after group formation, (3) following the introduction of females, and (4) in larger groups with more frequent dominance interactions. Thus, plasma cortisol levels may reflect what has been previously described as the "mean level of environmental stimulation" (Welch, 1964). This interpretation is not a new one since the responsiveness of the pituitary-adrenal system to psychological variables has already been documented in extensive review articles (Mason, 1968a; Hennessy and Levine, 1979; Rose, 1980). It is also noteworthy that dominant animals had higher levels of both cortisol and testosterone simultaneously because of the commonly reported inverse relationship between adrenal and gonadal output (Parkes and Deanesley, 1966; Mason, 1968b; Doerr and Pirke, 1976; Andrews, 1978). The stress-induced suppression of gonadal function which can occur in response to high levels of pituitary-adrenal hormones actually appears to be an accentuated physiological effect which differs from the ongoing endocrine relationships in undisturbed animals.

The tendency of dominant males to show higher plasma levels of cortisol may be partially explained by invoking a traditional view of dominance hierarchies which suggested that they can result in a differential "attention structure" (Chance, 1967). When monkeys are pair-housed, the subordinates may rely on the behavioral reactions of dominant males to assess the significance of environmental changes, and as a consequence, dominant subjects may stay in a more vigilant and aroused state. Similarly, dominant animals have been reported to show greater hormonal responsiveness than subordinates to changes in the social milieu (Manogue et al., 1975; Vogt et al., 1980). This relationship between dominance and adrenocortical secretion could shift, however, if the social or housing conditions are altered such that subordinates are now forced to continually avoid and anxiously orient towards dominant animals. In fact, we have now observed two groups of squirrel monkeys in which plasma cortisol levels were higher in subordinate animals (Vogt et al., 1980; Coe et al., in preparation). In both cases, the males were living under provocative conditions, either with females or visually exposed to them, and the groups were subjected to frequent environmental disturbances. In addition, the subjects of both groups were squirrel monkeys of Guyanese origin which tend to be more emotionally reactive in captivity (MacLean, 1964). Dominant males of the Guyanese variety also tend to harass subordinates and females more frequently after group formations (Travis and Holmes, 1974; Coe, personal observation).(3)

This shift in the effect of dominance across different social conditions may also be important for evaluating the relevance of the present findings to squirrel monkeys living under natural conditions. It is likely that hormone-behavior relationships would be most delineated when the number of animals in the troop is relatively small, such as in groups of 15-25 monkeys which usually have three to four adult males (Thorington, 1968), rather than in larger troops which can consist of more than 100 animals. Similarly, the effect of dominance on endocrine physiology would probably be greatest at times of higher inter-male competition, such as during the mating season when sexual excitement induces more frequent and intense dominance behavior, rather than during the birth season when males tend to be more passive and asocial (Baldwin, 1968).

This interaction between annual biorhythms and the effect of dominance was, in fact, one of the more intriguing findings of the current work. Squirrel monkeys have the capacity to undergo dramatic surges in hormone levels (Wilson et al., 1978), particularly during the mating season, and the effect of dominance was most manifest in terms of altering the frequency of these high surges. All males were able to show the full range in hormone titers but dominant animals produced these high surges more frequently, resulting in higher overall levels. The effect of dominance rank on testosterone levels in the four-male group was due primarily to the two higher-ranking animals showing these seasonal surges, whereas the hormone levels in the subordinate males remained lower and were more constant. A recent study we have conducted on diurnal rhythms has further emphasized this point in that the effect of dominance on adrenal and gonadal hormones was prominent only at times of the day when the hormone levels peaked (Coe, Gonzalez and Levine, in preparation).

An understanding of the relationship between behavioral and physiological processes necessitates considering a variety of factors simultaneously since endocrine activity reflects the confluence of diverse stimuli, and hormones subserve a number of functions in addition to their role in creating behavioral predispositions. Nevertheless, it is apparent that sociality is one of the most salient variables in a gregarious organism. As Sherrington (1933) aptly wrote in discussing the relationship between endogenous and exogenous processes, "The question who turns the key, to use that metaphor, is soon answered, the outside world."

ACKNOWLEDGEMENTS

The authors would like to acknowledge the help of Ms. Helen Hu and Ms. Brenda Siddall in running the radioimmunoassays. Special thanks are also due Ms. Edna Lowe, Mr. Edward Rich, Ms. Deborah Franklin, Ms. Cora Spaulding for their assistance in behavioral observations. Research support was provided by Grants No. MH-23645 from NIMH, HD-02881 from NICH&HD, and Research Scientist Award MH-19936 to Seymour Levine; MH-21178 from NIMH to Julian Davidson.

NOTES

1. The reader is referred to numerous studies on the behavior of squirrel monkeys for further details; Baldwin, 1968; Plotnik et al., 1968; Candland et al., 1973; Alvarez, 1975; Smith et al., 1977; Vaitl, 1977; Coe and Rosenblum, 1978; Mendoza et al., 1978b.

2. For further details of the blood sampling and assay procedures, see Coe et al., 1978; Mendoza et al., 1978a. Cortisol levels were assessed by the radioimmunoassay methods of Klemm and Gupta (1975), using antiserum F21-53 from Endocrine Sciences; testosterone levels were assessed by the radioimmunoassay methods of Frankel et al. (1975), using antiserum CEO #1 supplied by Dr. B. V. Caldwell.

3. The issue of subspecies differences goes beyond the scope of the present paper but, for those readers who might use Guyanese squirrel monkeys, it should be mentioned that they also have lower overall levels of steroid hormones which may be related to their smaller body size.

REFERENCES

Abbott, D.H. and Hearn, J.P. Physical, hormonal and behavioral aspects of sexual development in the marmoset monkey, Callithrix jacchus. J. Reprod. Fertil. 53, 155-166 (1978).

Alvarez, F. Social hierarchy under different criteria in groups of squirrel monkeys, Saimiri sciureus. Primates 16, 437-455 (1975).
Andrews, R.V. Influence of the adrenal gland on gonadal function, in Regulatory Mechanisms Affecting Gonadal Hormone Action, Advances in Sex Hormone Research, Vol.3, Thomas and Singhal, eds. University Park Press, Baltimore (1978), pp. 197-215.
Bailey, A. Personal communication (1975).
Baldwin, J.D. The social behavior of adult male squirrel monkeys (Saimiri sciureus) in a semi-natural environment. Folia primat. 9, 281-314 (1968).
Baldwin, J.D. Reproductive synchronization in squirrel monkeys (Saimiri). Primates 11, 317-326 (1970).
Bernstein, I.S. Primate status hierarchies, in Primate Behavior, Vol.1, Rosenblum, ed. Academic Press, New York (1970), pp. 71-109.
Bernstein, I.S., Gordon, T.P., and Peterson, M. Role behavior of an agonadal alpha-male rhesus monkey in a heterosexual group. Folia primat. 32, 263-267 (1979a).
Bernstein, I.S., Rose, R.M., and Gordon, T.P. Behavioral and hormonal responses of male rhesus monkeys introduced to females in the breeding and non-breeding seasons. Anim. Behav. 25, 609-614 (1977).
Bernstein, I.S., Rose, R.M., Gordon, T.P., and Grady, C.L. Agonistic rank, aggression, social context and testosterone in male pigtail monkeys. Aggressive Behav. 5, 329-339 (1979b).
Brown, G.M., Grota, L.J., Penny, D.P., and Reichlin, S. Pituitary-adrenal function in the squirrel monkey. Endocrinology 86, 519-529 (1970).
Candland, D.K., Dresdale, L., Leiphart, J., Bryan, D., Johnson, C., and Nazar, B. Social structure of the squirrel monkey (Saimiri sciureus, Iquitos); relationships among behavior, heart-rate, and physical distance. Folia primat. 20, 211-240 (1973).
Candland, D.K. and Leshner, A.I. A model of agonistic behavior: endocrine and autonomic correlates, in Limbic and Autonomic Systems Research, DiCara, ed. Plenum Press, New York (1974), pp. 137-163.

Chamove, A.S. and Bowman, R.E. Rhesus plasma cortisol response at four dominance positions. Aggressive Behav. 4, 43-55 (1978).

Chance, M.R.A. Attention structure as the basis of primate rank orders. Man 2, 503-518 (1967).

Chen, J.J., Smith, E.R., Gray, G.D., and Davidson, J.M. Seasonal changes in plasma testosterone and ejaculatory capacity in squirrel monkeys (Saimiri sciureus). Primates 22, 253-260 (1981).

Christian, J.J. Seasonal changes in the adrenal glands of woodchucks (Marmota monax). Endocrinology 71, 431-447 (1962).

Chrousos, G.P., Renquist, D., Brandon, D., Eil, C., Pugeat, M., Cutler, G.B., Loriaux, D.L., and Lipsett, M.B. Glucocorticoid hormone "resistance" in two primate species. Clin. Res. 29, 504A (1981).

Clark, L.D. and Nakashima, E.W. Measurement of social dominance in squirrel monkeys. Behav. Res. Meth. Instru. 4, 143-144 (1972).

Coe, C.L., Gonzalez, C.A., and Levine, S. Diurnal rhythms in the adrenal and gonadal hormones of the squirrel monkey (in preparation).

Coe, C.L. and Levine, S. Psychoendocrine relationships underlying reproductive behavior in the squirrel monkey. Int'l. J. Ment. Health 10, 22-42 (1981).

Coe, C.L., Mendoza, S.P., Davidson, J.M., Smith, E.R., Dallman, M.F., and Levine, S. Hormonal response to stress in the squirrel monkey (Saimiri sciureus). Neuroendocrinology 26, 367-377 (1978).

Coe, C.L., Mendoza, S.P., and Levine, S. Social status constrains the stress response in the squirrel monkey. Physiol. Behav. 23, 633-638 (1979).

Coe, C.L. and Rosenblum, L.A. Annual reproductive strategy of the squirrel monkey (Saimiri sciureus). Folia primat. 29, 19-42 (1978).

Deag, J.M. Aggression and submission in monkey societies. Anim. Behav. 25, 465-474 (1977).

Dixson, A.F. Androgens and aggressive behavior in primates: a review. Aggressive Behav. 6, 37-67 (1980).

Dixson, A.F., Martin, R.D., Bonney, R.C., and Fleming, D. Reproductive biology of the owl monkey, Aotus frivirgatus griseimembra, in Non-Human Primate Models for Study of Human Reproduction, Anand Kumar, ed. Karger, Basel (1980), pp. 61-68.

Doerr, P. and Pirke, K.M. Cortisol-induced suppression of plasma testosterone in normal adult males. J. Clin. Endocrin. Metab. 43, 622-629 (1976).

Eaton, G.G. and Resko, J.A. Plasma testosterone and male dominance in a Japanese macaque (Macaca fuscata) troop compared with repeated measures of testosterone in laboratory males. Horm. Behav. 5, 251-259 (1974).

Epple, G. Lack of effects of castration on scent marking, displays and aggression in a South American primate (Saguinus fuscicollis) Horm. Behav. 11: 139-150 (1978).

Frankel, A.I., Mock, E.J., Wright, W.W., and Kamel, F. Characterization and physiological validation of radioimmunoassay for plasma testosterone in the male rat. Steroids 25, 73-98 (1975).

Gartlan, J.S. Structure and function in primate society. Folia primat. 8, 89-120 (1968).

Gordon, T.P., Rose, R.M., Grady, C.L., and Bernstein, I.S. Effects of increased testosterone secretion in the behavior of adult male rhesus living in a social group. Folia primat. 32, 149-160 (1979).

Green, R., Whalen, R.E., Rutley, B., and Battie, C. Dominance hierarchy in squirrel monkeys (Saimiri sciureus). Folia primat. 18, 185-195 (1972).

Hall, K.R.L. Social organization of the Old World monkeys and apes. Symp. Zool. Soc. Lond. 14, 265-289 (1965).

Hennessy, J.W. and Levine, S. Stress, arousal and the pituitary-adrenal system: a psychoendocrine model, in Progress in Psychobiology and Physiological Psychology, Vol. 8, Sprague and Epstein, ed. Academic Press, New York (1979), pp. 133-178.

Hennessy, M.B., Mendoza, S.P., Coe, C.L., Lowe, E.L., and Levine, S. Androgen-related behavior in the squirrel monkey; an issue that is nothing to sneeze at. Behav. Neur. Biol. 30, 103-108 (1980).

Keverne, E.B. Sexual and aggressive behavior in social groups of talapoin monkeys, in Sex, Hormones and Behavior, Ciba Foundation Symposium 62. Excerpta Medica, Amsterdam (1979), pp. 271-287.

Klemm, W. and Gupta, D. A routine method for the radioimmunoassay of plasma cortisol without chromatography, in Radioimmunoassay of Steroid Hormones, D. Gupta, ed. Verlag Chemie, Weinheim, Germany (1975), pp. 143-151.

van Kreveld, D. A selective review of dominance-subordination relations in animals. Genet. Psychol. Monogr. 81, 141-173 (1970).

Liptrap, R.M. and Raeside, J.I. A relationship between plasma concentrations of testosterone and corticosteroids during sexual and aggressive behaviour in the boar. J. Endocrin. 76, 75-85 (1978).

MacLean, P.D. Mirror display in the squirrel monkey (Saimiri sciureus). Science 146, 950-952 (1964).

Manogue, K.R., Leshner, A.I., and Candland, D.K. Dominance status and adrenocorticol reactivity to stress in squirrel monkeys (Saimiri sciureus). Primates 16, 457-463 (1975).

Mason, J.W. A review of psychoendocrine research on the pituitary-adrenal cortical system. Psychosom. Med. 30, 576-607 (1968a).

Mason, J.W. Organization of the multiple endocrine responses to avoidance conditioning in the monkey. Psychosom. Med. 30, 774-790 (1968b).

Mendoza, S.P., Coe, C.L., Lowe, E.L., and Levine, S. The physiological response to group formation in adult male squirrel monkeys. Psychoneuroendocrinology 3, 221-229 (1979).

Mendoza, S.P., Lowe, E.L., Davidson, J.M., and Levine, S. Annual cyclicity in the squirrel monkey (Saimiri sciureus): the relationship between testosterone, fatting and sexual behavior. Horm. Behav. 11, 295-303 (1978a).

Mendoza, S.P., Lowe, E.L., and Levine S. Social organization and social behavior in two subspecies of squirrel monkeys (Saimiri sciureus). Folia primat. 30, 126-144 (1978b).

Mendoza, S.P., Lowe, E.L., Resko, J.A., and Levine S. Seasonal variations in gonadal hormones and social behavior in squirrel monkeys. Physiol. Behav. 20, 515-522 (1978c).

Nagle, C.A., Denari, J.H., Riarte, A., Quiroga, S., Zarate, R., Germino, N.I., Merlo, A., and Rosner, J.M. Endocrine and morphological aspects of the menstrual cycle in the cebus monkey (Cebus appela), in Non-Human Primate Models for Study of Human Reproduction, Anand Kumar, ed. Karger, Basel (1980), pp. 69-81.

Nieschlag, E. and Wickings, E.J. Testicular and adrenal steroids in the adult rhesus monkey, in Non-Human Primate Models for Study of Human Reproduction, Anand Kumar, ed. Karger, Basel (1980), pp. 136-147.

Parkes, A.S. and Deanesley, R. Relations between the gonads and the adrenal glands. Marshall's Physiology of Reproduction, Vol. 3, A.S. Parkes, ed. Little, Brown & Co., Boston (1966), pp. 1064-1111.

Ploog, D.W., Blitz, J., and Ploog, F. Studies on social and sexual behavior of the squirrel monkey (Saimiri sciureus). Folia primat. 1, 29-66 (1963).

Plotnik, R., King, F.A., and Roberts, L. Effects of competition on the aggressive behavior of squirrel and Cebas monkeys. Behaviour 32, 315-322 (1968).

Rose, R.M. Endocrine responses to stressful psychological events. Psychiat. Clin. North America 3, 251-275 (1980).

Rose, R.M., Bernstein, I.S., Gordon, T.P., and Catlin, S.F. Androgens and aggression; a review and recent findings in primates, in Primate Aggression, Territoriality and Xenophobia, Holloway, ed. Academic Press, New York (1974), pp. 275-304.

Rosenblum, L.A. Mother-infant relations and early behavioral development in the squirrel monkey, in The Squirrel Monkey, Rosenblum and Cooper, eds. Academic Press, New York (1968), pp. 207-233.

Rowell, T.E. The concept of social dominance. Behav. Biol. 11, 131-154 (1974).

Sassenrath, E.N. Increased adrenal responsiveness related to social stress in rhesus monkeys. Horm. Behav. 1, 283-298 (1970).

Sherrington, C.S. The Brain and its Mechanisms. Cambridge University Press, Cambridge (1933).

Smith, M., Harris, R.J., and Strayer, F.F. Laboratory methods for the assessment of social dominance among captive squirrel monkeys. Primates 18, 977-984 (1977).

Thorington, R.W., Jr. Observations of squirrel monkeys in a Colombian forest, in The Squirrel Monkey, Rosenblum and Cooper, eds. Academic Press, New York (1968), pp. 69-85.

Travis, J.C. and Holmes, W.N. Some physiological and behavioral changes associated with oestrus and pregnancy in the squirrel monkey. J. Zool., Lond. 174, 41-66 (1974).

Vaitl, E.A. Social context as a structuring mechanism in captive groups of squirrel monkeys (Saimiri sciureus). Primates 18, 861-874 (1977).

Vogt, J.L., Coe, C.L., Lowe, E.L., and Levine, S. Behavioral and pituitary-adrenal response of adult squirrel monkeys to mother-infant separation. Psychoneuroendocrinology 5, 181-190 (1980).

Welch, B.L. Psychophysiological response to the mean level of environmental stimulation: a theory of environmental integration, in Symposium on Medical Aspects of Stress in the Military Climate, April 1964. Walter Reed Army Institute of Research, Washington, D.C. (1964), pp. 39-95.

Wilson, M.I., Brown, G.M., and Wilson, D. Annual and diurnal changes in plasma androgen and cortisol in adult male squirrel monkeys (Saimiri sciureus) studied longitudinally. Acta Endocrin. 87, 424-433 (1978).

Wolf, R.C., O'Connor, R.F., and Robinson, J.A. Cyclic changes in plasma progestins and estrogens in squirrel monkeys. Biol. Reprod. 17, 228-231 (1977).

Chapter 2
Plasma Testosterone, Sexual and Aggressive Behavior in Social Groups of Talapoin Monkeys

E.B. Keverne, J.A. Eberhart, and R.E. Meller

INTRODUCTION

It has long been known in non-primate species that social stimuli may modify the hormonal state of an individual (Matthews, 1938; Guhl, 1961). More specifically, in mice the presence of females may increase male testosterone levels (Macrides et al., 1975), while on the other hand social stress can suppress gonadotrophin secretion (Bronson, 1973; Bronson and Eleftheriou, 1965).

Attention has also been given to the possibility of such effects in primates. The pioneering work of Mason and his co-workers on adult male rhesus monkeys (Mason and Brady, 1965; Mason, 1968) showed that a wide variety of endocrine secretions were responsive to psychological stimuli. Since these early laboratory studies, long-term investigations have been carried out on large groups of rhesus monkeys (reviewed in Bernstein et al., 1974; Rose et al., 1974), in which hormonal responses of individuals have been monitored during the course of various social manipulations. It was found that testosterone levels in adult males rose following their introduction to oestrous females (Rose et al., 1972); in contrast, when males were introduced singly into well-established groups, whose members attacked and defeated them, their testosterone levels fell dramatically (Rose et al., 1975). Further, the victorious males in these groups experienced an increase in plasma testosterone levels.

Such findings clearly have significance for an understanding of the relationship between social dominance and testosterone secretion in non-human primates. It is perhaps surprising then that this relationship is still unclear: some reports provide evidence for a positive correlation between male rank and testosterone levels (Rose et al., 1971), while others fail to do so (Eaton and Resko, 1974). However these studies simply make statements concerning rank and hormone levels, rather than considering the dynamics of behavioural interaction that occur in such social situations. It is our view that a detailed analysis of the behavioural situation is essential, with the significance of dominance being formulated in terms of the sexual as well as the aggressive activity of the individual. This approach is likely to be more productive than one in which dominance is referred to as if it were a unitary quality definable independently of the individual's behaviour. The present paper then is the result of a long-term study on three social groups of adult male and female talapoin monkeys, in which the relationship between male status and testosterone secretion is analysed in terms of observed behavioural events. In these experiments the social structure of each group and the hormonal state of the females were manipulated, and both the behavioural and endocrine responses of individuals were monitored.

MATERIALS AND METHODS

Twelve adult male, and nine adult female talapoin monkeys (<u>Miopithecus talapoin</u>) were used. Two of the males in one group had been castrated and all females had been ovariectomised at least three years previously. Castrated males received testosterone replacement in the form of silastic implants of crystalline hormone (maintaining testosterone levels of 8-10 ng/ml) throughout the study. Females were given oestradiol implants (releasing 2-3 µg/24 hours) at intervals during the study, to mimic follicular phase oestrogen levels. Behavioural data from castrated males are not included in this analysis.

Individuals whose behaviour was being observed were housed in groups of four males and four or five females in large (5' x 8' x 11') group cages (see Keverne et al., 1978, for description). The details of behavioural observations are given elsewhere (Dixson et al.,, 1975); in brief, sexual aggressive and social interactions were recorded, with all animals being monitored continuously.

In group I, males and females had continuous access to one another, while in Groups II and III sexes were housed in the group cage, but were partitioned from each other except during the test periods. Animals were observed for two 50 minute periods (at 0800 or 0900 and 1700) per day for a minimum of ten days during each treatment. Animals not being observed were housed singly in a separate room.

The following manipulations were performed:

I. For Groups II and III all males remained in the large cage, but were partitioned from females when the females were
   A. Untreated with oestrogen, and
   B. Oestrogen-treated (i.e., sexually attractive).

II. For all three groups each male was
   A. In the mixed-sex social group: females were oestrogen-treated and other males were present,
   B. With the same oestrogen-treated females in the absence of other males, and
   C. In isolation: caged singly in a separate room.

Dominance rank was determined by the direction of aggression between animals when all males were housed with oestrogen-treated females. Data in Table 1 represent a minimum of 1000 minutes of observation. For each group, males were ranked in order of threats and attacks given to, and received from, each other male. Although there were some changes in rank (see Results), dominance hierarchies remained relatively stable throughout the experiment (up to three years).

Twice weekly, immediately following the 1700 hour observation period, blood samples (2-4 ml) were taken by femoral venipuncture under ketamine anaesthesia. Plasma was stored at $-20^\circ C$ until testosterone determination by radioimmunoassay procedures described elsewhere (Keverne et al., 1978). Behavioural observations were not made on the day following blood sampling.

STATISTICAL ANALYSIS

A two-way (Animal X Treatment) analysis of variance (SPSS: Nie et al., 1975) was run on testosterone data of all males of each group for three conditions (Isolation, All males with females, All males with oestrogen-treated females). This analysis revealed significant main effects of Condition in Groups I and II ($F = 18.127$, $p < .001$), of Rank in all Groups ($F = 8.677$, $p < .001$; $F = 22.067$, $p < .001$; $F = 2.920$, $p < .05$), and of Condition X Rank Interaction in Groups II and III ($F = 11.349$, $p < .001$; $F = 3.794$, $p < .005$). Since differences in testosterone were rank-related, student's t-tests (Moore et al., 1972) were performed on testosterone data from the most dominant and most subordinate males, as these males most clearly demonstrate the effects of the hierarchy. Behavioural data (not assumed to be normally distributed) were compared with Mann-Whitney U tests (Siegel, 1956).

RESULTS

Among the males of each group there was a clear linear dominance hierarchy, as determined by the direction of aggression between individuals. Table 1 illustrated that in Groups I and II, the direction of aggression between males of different rank was entirely unilateral (from more dominant to more subordinate animals). Although subordinate males in Group III occasionally threatened or attacked dominant males, this aggression was never initiated by subordinates, but occurred in response to aggression received, and was infrequent in comparison to received aggression. In any event, these data clearly indicate a linear dominance hierarchy.

TABLE 1

Determination of male dominance rank by the direction of aggression. See text for details.

A. *Group I*   Attacks (Threats) Given

Received \ Male	M	C1	C2	330
M	—			
C1	7 (17)	—		
C2	6 (11)	3 (0)	—	
330	12 (17)	20 (1)	5 (3)	—

B. *Group II*   Attacks (Threats) Given

Received \ Male	D	Fo	J5	Sh
D	—			
Fo	15 (1)	—		
J5	2 (1)	9 (3)	—	
Sh	3 (1)	7 (1)	35 (3)	—

C. *Group III*   Attacks (Threats) Given

Received \ Male	S	I	D	N
S	—		(3)	15 (1)
I	2 (5)	—		7 (8)
D	12 (10)	18 (10)	—	3 (2)
N	60 (13)	308 (59)	628 (85)	—

Unless otherwise indicated, results are presented for only the most dominant and the most subordinate males; these highest and lowest ranking males are hereafter referred to simply as "dominant" and "subordinate" males.

Effects of Housing Conditions on Plasma Testosterone

When the males were out of the mixed-sex social group and caged singly, there was no difference in plasma testosterone levels between dominant males and subordinate males. In Groups I and II testosterone levels were similar in dominant and subordinate males, and in Group III testosterone was higher in the subordinate male than in the dominant ($t = 5.47$, $p < 0.001$). When all males were in the social group and had access to oestrogen-treated females, however, the dominant males had higher mean plasma testosterone levels than did subordinate males (Group I $t = 2.36$, $p < 0.05$; Group II $t = 3.29$, $p < 0.001$; Group III $t = 3.05$, $p < 0.001$). Thus, isolation levels of testosterone could not be used to predict either dominance rank or testosterone level in the social group. The most significant findings, therefore, was the increase in plasma testosterone which occurred in the dominant male with access to attractive females (Group I $t = 4.6$, $p < 0.001$; Group II $t = 3.83$, $p < 0.001$; Group III $t = 6.19$, $p < 0.001$), an increase which failed to occur in subordinates (Fig. 1).

Effects of the Presence of Attractive Females on Behaviour and Plasma Testosterone

Treating the females with oestrogen but not permitting males access to them had no significant effect on male aggressive or sexual behaviour when compared to the untreated/no access condition. Plasma testosterone levels increased in some males following oestrogen administration, but this occurred in subordinate as well as in dominant males. When access to attractive females was

SEX, AGGRESSION, AND PLASMA TESTOSTERONE 39

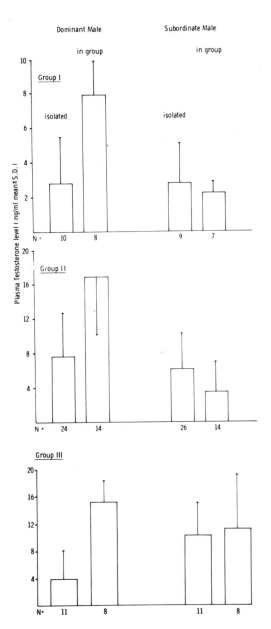

Figure 1. Mean plasma testosterone (ng ml^{-1} + S.D.) of dominant and subordinate males when caged singly or in the social group with ovariectomized, oestrogen-treated females. Number of samples given by N. For significance values see text. Samples were collected weekly during a 10-12 week period in isolation and a 6-8 week period in the social group.

permitted, however, consistent and marked differences in behavioural and endocrine responses were seen between dominant and subordinate males. Only dominant males showed high levels of sexual activity: they mounted and ejaculated with females, who directed many sexual invitations to them. Subordinate males, in contrast, rarely mounted and never ejaculated, despite showing sexual interest in females (i.e., occasional inspections of females' perinea); these males were rarely solicited. Dominant males were also more aggressive than previously, while subordinates received more aggression at this time. With access to oestrogen-treated females, dominant males experienced significant increases in plasma testosterone (Group II $t = 6.28$, $p < 0.001$; Group III $t = 6.19$, $p < 0.001$) while levels of this hormone fell in subordinates.

Rank Change and Plasma Testosterone

During the course of four years' observations on three groups of monkeys, there were eight cases of an intact male changing rank (four rises and four falls in the hierarchy). All such changes occurred within the first six months of group formation, and five were associated with oestrogen-treatment of the females. Rising in rank was twice accompanied by a significant increase in testosterone levels (Group I $t = 3.14$, $p < 0.01$; Group II $t = 2.60$, $p < 0.01$), and twice with no change in this hormone ($p = n.s.$). Three of four decreases in rank were accompanied by a decrement in testosterone (Group III $t = 4.32$, $p < 0.001$; Group II $t = 3.68$, $p < 0.001$), and one was associated with no change. Significantly, rising in rank was never associated with a decrease in testosterone, nor was falling in rank associated with an increase. When there was a change in plasma testosterone following an increase in rank, levels of testosterone were higher after the change, with the converse following a decrease in rank.

Aggressive Behaviour and Plasma Testosterone

A question of some importance is why plasma testosterone increases only in the dominant male, and not in the subordinate, when the females received oestrogen, and whether this change is associated primarily with aggressive or sexual behaviour.

a) Aggression given

The display of aggression and its effects on plasma testosterone, independent of sexual activity, could be assessed in both the all-male group and the mixed-sex group when females were not treated with oestradiol. In the all-male group dominant individuals attacked and threatened other males, yet their plasma testosterone levels were not significantly different from when these males were in isolation (Fig. 2). Moreover, in the all-male group when access to females was not allowed there was no significant difference in plasma testosterone levels between dominant and subordinate males ($p = n.s.$) although dominant males showed aggression and subordinates did not.

In the mixed-sex group prior to females receiving oestrogen the dominant male showed aggressive, but no sexual, behaviour (Fig. 2 Group II). Plasma testosterone in the dominant male at this time was not different from that of the subordinate ($p = n.s.$) but it did increase when the dominant male became sexually active ($t = 3.38$, $p < 0.001$). Furthermore, a low-ranking male (rank 3) displayed more aggression ($U = 5.94$, $p < 0.001$) without becoming sexually active when the females received oestradiol, but his plasma testosterone levels did not change ($p = n.s.$, Fig. 2, Group III). Becoming dominant *per se* may, however, be associated with an increase in plasma testosterone even in an all-male group where heterosexual behaviour is absent. Although this change in rank is associated with an increase in the display of aggression, it is also associated with a decrease in aggression received, and which of these factors is influencing testosterone cannot be determined from these data.

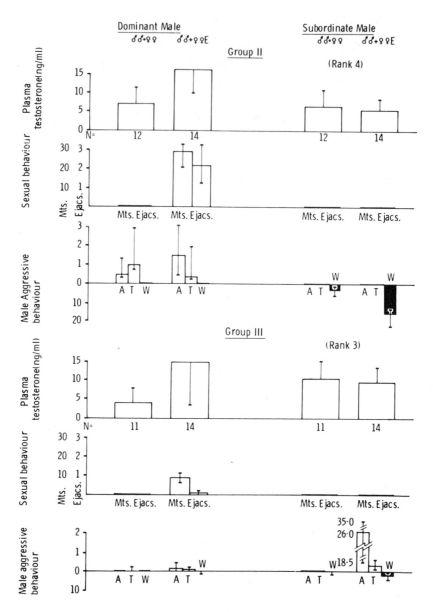

Figure 2. Changes in plasma testosterone (ng^{-1} + S.D.) in dominant and subordinate males in relationship to sexual and aggressive interactions. 1st Column: females unreated; 2nd Column: females treated with oestrogen. Behavioural data shown as median score/100 mins plus IQR. A=attacks, T=threats, W=withdrawals.

b) Aggression received

In an all-male group, one animal experienced a decrease in rank and a concomitant increase in aggression received, corresponding with a decrease in plasma testosterone in the absence of sexual behaviour. This decrease was not significant; however, two subordinates did exhibit significant decreases in this hormone on being given access (with other males) to oestrogen-treated females (Group II $t = 2.67$, $p < 0.02$; Group III $t = 5.52$, $p < 0.001$). When access to attractive females was permitted at this time, subordinates showed no sexual behaviour but received significantly more aggression from higher-ranking males (Group II $U = 13$, $p < 0.01$; Group III $U = 9.0$, $p < 0.007$) than they had prior to oestrogen-treatment of the females. The significant decrease in plasma testosterone was thus associated with the receipt of increased aggression.

Since female talapoin monkeys can, and frequently do, dominate males, aggression received from females may be important in some situations. In the absence of any male cagemates, a male may receive markedly more aggression from previously unaggressive females. This occurred in the case of the dominant male of Group I. On removal of this male's subordinates, females became more aggressive towards him (attacks: $U = 19$, $p < 0.02$; threats: $U = 7$, $p < 0.0002$) and his testosterone levels fell ($t = 3$, $p < 0.01$, Fig. 3). Decreased testosterone and the receipt of increased aggression was associated in this male with total inhibition of his sexual activity. In Group II, aggression from females remained low, sexual activity was still observed, and testosterone levels in the dominant male did not change (Fig. 3).

Sexual Behaviour and Plasma Testosterone

In order to determine if sexual behaviour per se influences testosterone secretion, changes in sexual interaction must be considered in situations where aggressive behaviour does not change. That the display of sexual activity is associated with

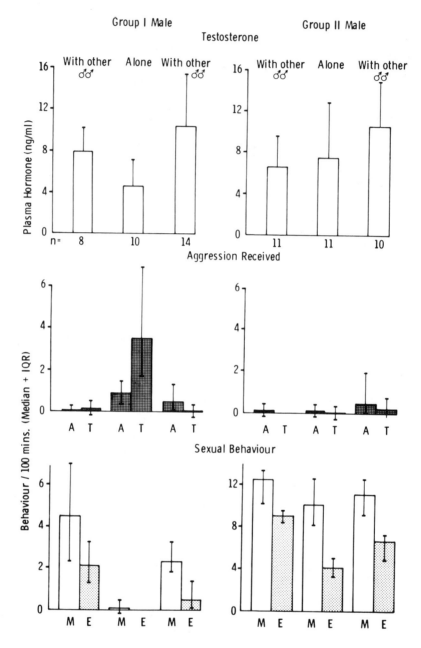

Figure 3. Effects of behavioural interactions with females on plasma testosterone (ng^{-1} + S.D.) of dominant males in the presence (1st and 3rd columns) or absence (middle column) of other males. Behavioural data shown as median score/100 mins and IQR. M=mounts, E=ejaculations. For significance values, see text.

an increase in testosterone in the absence of altered aggressive behaviour has been shown in the case of the dominant males who were permitted access to attractive (oestrogen-treated) females (Fig. 2). In dominant males of both groups, significant increases in plasma testosterone (Group II $t = 3.38$, $p < 0.001$, Group III $t = 6.2$, $p < 0.001$) were associated with increases in sexual activity, while the subordinate males showed no sexual behaviour and their testosterone levels remained low.

Further support for this association is seen in another situation, namely when each male of Group II was introduced to the group of attractive females in the absence of other males. As described above, the formerly dominant male of Group II continued to show sexual behaviour (Fig. 4). In this situation the subordinate male also showed sexual behaviour, mounting and ejaculating for the first time with the females (Fig. 4). In this group aggressive interactions between the sexes were relatively infrequent. Strikingly, at this time both males showed significantly higher levels of testosterone than when out of the social group (dominant $t = 2.21$, $p < 0.001$; subordinate $t = 5.42$, $p < 0.001$) (Fig. 4), suggesting that when aggression is unchanged, sexual activity can lead to increased testosterone levels.

Plasma Testosterone, Sexual Behaviour and Female Proceptivity

Females in each group directed significantly more sexual invitations to dominant males than to subordinates (Fig. 5). Subordinate males received virtually no sexual invitations and showed no sexual behaviour; their plasma testosterone levels were also low at this time. Despite the marked difference in sexual activity of the dominant males of each group, female sexual invitations were very high to each of these males. Since dominant males were sexually active and subordinates were not, it could be argued that sexual invitations followed males' sexual behaviour. Against such a view, however, is the observation that the dominant male showing

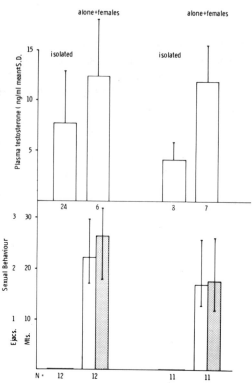

Figure 4. Mean plasma testosterone (ng^{-1} + S.D.) and sexual behaviour of dominant and subordinate males when caged singly or alone with oestrogen-treated females. Behavioural data are given as median scores/100 mins + IQR. No. of plasma samples under upper column.

fewest ejaculations received the highest number of female sexual invitations (Fig. 5).

DISCUSSION

The results of this long-term study support the view that the interaction beteen hormones and behaviour is such that social events may influence endocrine state as well as vice versa (Young, 1961; Beach, 1965). Thus, when male talapoins were introduced to attractive females, plasma testosterone levels increased in dominant but not in subordinate males. Changes in social rank also influence plasma testosterone, with increases in

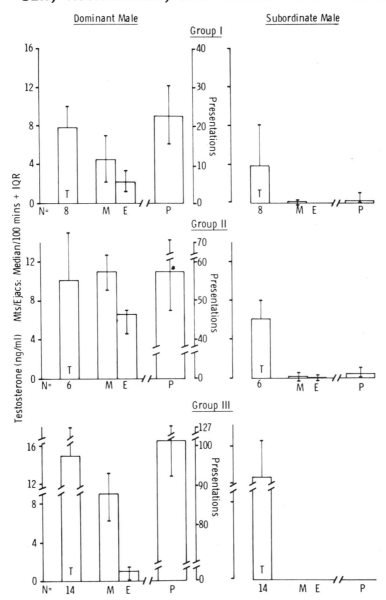

Figure 5. Plasma testosterone, sexual interactions and female proceptivity. Significantly higher levels of sexual invitations are directed towards the dominant male and not the subordinate. The dominant male also shows higher levels of sexual behaviour and plasma testosterone. See text for significance values. M=mounts, E=ejaculations, P=female presentation.

testosterone accompanying increases in rank in some cases, and the converse for decreases in dominance rank. Since dominance was defined in terms of the direction of aggression, it might be anticipated that social events influencing plasma testosterone levels can be analysed simply by considering aggressive behaviour.

This proved not to be the case, since in the presence of attractive females, changes in testosterone were associated with changes in a number of behavioural interactions. It is therefore important to establish which of these behaviours are the prime movers influencing testosterone changes: is it the reaction to females and sexual experiences with them, or is it aggressive behaviour either given or received which influences endocrine state?

Considering the relationship between rank and testosterone, a careful analysis of the data from these groups of monkeys suggests that, although dominance and raised testosterone levels are positively correlated, the giving of aggression _per se_ is without effect on this hormone. While it appears that receiving aggression is directly associated with decreased testosterone, it could be argued that receipt of aggression affects testosterone indirectly, by inhibiting sexual activity. However, decreased testosterone levels were seen independently of changes in the individual's sexual activity, but at a time when the male received increased aggression. In another situation a possibly indirect effect of aggression on decreased testosterone was observed: when the dominant male of Group I was alone with females, the latter increased their aggression toward him and at this time he showed no sexual behaviour. The summation of the direct and the indirect routes of testosterone suppression by aggression would thus seem to be very effective. It has also been shown here that the reduction of aggression received (brought about by removing dominant males) enabled a former subordinate male to show sexual behaviour, which was accompanied by a rise in testosterone. Since he exhibited the same increase in hormone levels when able to copulate with attractive females as did the dominant male, the subordinate's endocrine system was potentially as responsive as that of the dominant male.

It is important to stress that in long-established groups aggressive interactions decline markedly in frequency, making the day-to-day determination of dominance from aggressive behaviour extremely difficult. Similar observations on free-ranging monkeys have led field workers to question the concept of dominance, suggesting that it may be an artifact of laboratory conditions (Rowell, 1967; Garlan, 1968). However, the present study shows that even in the absence of behaviours permitting the immediate determination of dominance, animals are aware of the hierarchy (see also Keverne, et al., 1978a); furthermore, this dominance hierarchy is reflected in markedly different endocrine profiles in individuals of different ranks. Thus, while dominant males generally have high testosterone, low cortisol, and low prolactin, subordinates typically have low testosterone, high cortisol, and high prolactin (Keverne, 1979). If the concept of dominance were to be broadened to include not only behavioural but also endocrine characteristics, its use and measurement might become less controversial and more meaningful.

A question of some importance is the biological significance of elevated plasma testosterone in dominant males and suppressed levels in subordinates. Since testosterone may facilitate aggressive behaviour in human and non-human primates (Heller and Nelson, 1945; Dixson and Herbert, 1977a), the association of high rank with high testosterone would reinforce the structure of the hierarchy. Testosterone does not appear to determine whether or not a male is sexually active: male primates continue to show sexual interest and behaviour long after castration has rendered plasma testosterone undetectable (Wilson et al., 1972; Phoenix et al., 1973). Further, the administration of testosterone to subordinate male talapoins in a social group fails to increase their sexual activity (Dixson and Herbert, 1977b). Moreover, subordinate males with low testosterone levels are also sexually interested in females, judging from their inspections of the females' perineal region. Yet, females show little interest in subordinate males, directing their sexual invitations predominantly toward dominant males. In a primate species (such as the talapoin monkey) in which females may dominate males (Dixson and Herbert, 1977a; Wolfheim and Rowell, 1972) this selective directioning of proceptive behaviour might be particularly important. The possibility, therefore, that high plasma testosterone levels are associated with increased male attractiveness cannot be ignored. In behavioural terms, high testosterone may facilitate male sexual performance (Phoenix et al., 1973; Michael and Wilson, 1974) and thereby increase attractiveness, although there is no evidence yet concerning such a phenomenon in the talapoin monkey. In reproductive terms, high levels of testosterone are known to be important in initiating and maintaining spermatogenesis (Steinberger, 1971).

Although the interrelations of testosterone, sexual activity and aggression are not yet fully understood, it is clear that in laboratory groups of talapoin monkeys these three factors serve to increase the reproductive potential of the dominant male and to decrease that of the subordinate. Contrary to reports in some other primate species in which subordinate males mate successfully (although less frequently than dominant males) (Sugiyama, 1971; Eaton, 1976), the distribution of sexual behaviour among captive talapoin males is particularly restricted (Dixson and Herbert, 1977b). Whether or not mating is equally restricted in the wild is difficult to assess. Because of the small size of talapoins (generally 0.75-1.5 kg), and their preference for densely-foliated forest canopy, detailed observation of natural behaviour is particularly difficult (Rowell, 1973; Wolfheim and Rowell, 1972, Rowell and Dixson, 1975). Data from the natural habitat suggest that large (N-70) social groups contain more than one sexually active male (Rowell and Dixson, 1975). Observation conditions in the latter study did not allow the identification of animals by rank. Yet, these authors "had the impression that copulations usually occurred at a distance from other males, and that males were tensely aware of other males during courtship and broke off interactions if other males approached" (p. 428). They also reported seeing "a male observing courtship and copulation of another male with apparent indifference" (p. 428). Although alternate explanations are possible, one interpretation of these data is that the highest-ranking males copulate with impunity in the presence of subordinate males, whereas the more numerous subordinate males seek to copulate in visual isolation from other males. This pattern of rank-differentiated mating strategies has been observed both in captive talapoin groups in this laboratory and in free-rangning hamadryas baboons (Kummer, 1968). Thus, although there are many differences between laboratory conditions and the natural habitat, the distribution of male sexual behaviour involves some form of regulation by other animals in both settings, and the effects demonstrated in the laboratory may well reflect the same mechanisms operating in the wild. In any event, these laboratory studies clearly show that social factors can have a marked influence on both behavioural and hormonal states.

Certainly, subordinate males in laboratory groups have the potential to mate, as these studies have illustrated, although this sexual activity was at a substantially lower level than in the dominant male in the same situation. Even optimising conditions for subordinates to exhibit sexual behaviour (by removing other males) does not, therefore, necessarily result in the same level of sexual activity as shown by individuals that have been dominant. This suggests that the effects of past social subordination can influence an individual's behaviour even in the absence of subordinating factors. It is clearly of biological significance that social constraints exert such a powerful influence on an individual's reproductive capabilities and that behavioural and endocrine mechanisms in this respect subserve the same end. Social rank in primates has long been considered an important influence on sexual success (Carpenter, 1942; Kaufman, 1965): the present study supports this view by demonstrating that social factors can affect such success by a direct influence on the endocrine state of the individual.

ACKNOWLEDGEMENTS

This work was supported by an MRC programme grant. We thank Gerald Moore for statistical analyses, Raith Overhill for drawing the figures, and Susan Currie for typing the manuscript.

J.A. Eberhart was supported by a Marshall Scholarship and Rachel E. Meller by a research fellowship from the Mental Health Foundation.

REFERENCES

Beach, F.A. Retrospect and prospect, in Sex and Behaviour, F.A. Beach, ed. J. Wiley and Sons, New York (1965), pp. 535-569.

Bernstein, I.S., Rose, R.M., and Gordon, T.P. Behavioural and environmental events influencing primate testosterone levels. J. Hum. Evol. 3, 517-525 (1974).

Bernstein, I.S., Gordon, T.P., Rose, R.M., and Paterson, M.S. Influences of sexual and social stimuli upon circulating levels of testosterone in male pigtail macaques. Behav. Biol. 24, 400-404 (1978).

Bronson, F.H. Establishment of social rank among grouped male mice: relative effects on circulating FSH, LH and corticosterone, Physiol. Behav. 10, 947-951 (1973).

Bronson, F.J. and Eleftheriou, B.E. Adrenal response to fighting in mice: separation of physical and psychological causes. Science 147, 627-628 (1965).

Carpenter, C.R. Sexual behaviour of free-ranging rhesus monkeys. J. Comp. Psychol. 33, 113-162 (1942).

Christian, J.J. Effect of population size on the adrenal glands and reproductive organs of male mice in populations of fixed size. Am. J. Physiol. 182, 292-300 (1955).

Christian, J.J. Endocrine adaptive mechanisms and the physiological regulation of population growth, in Physiological Mammalogy, W. Mayer and R. van Gelder, eds. Academic Press, New York (1963).

Dixson, A.F. and Herbert, J. Testosterone, aggressive behaviour and dominance rank in captive adult male talapoin monkeys (Miopithecus talapoin). Physiol. Behav. 18, 539-543 (1977a).

Dixson, A.F. and Herbert, J. Gonadal hormones and sexual behaviour in groups of adult talapoin monkeys (Miopithecus talapoin). Horm. Behav. 8, 141-154 (1977b).

Dixson, A.F., Scruton, D.M., and Herbert, J. Behaviour of the talapoin monkey (Miopithecus talapoin) studied in groups, in the laboratory. J. Zool., Lond. 176, 177-210 (1975).

Eaton, G.G. The social structure order of Japanese macaques. Sci. Amer. 253, 96-106 (1976).

Eaton, G.G. and Resko, J.A. Plasma testosterone and male dominance in a Japanese macaque (Macaca fuscata) troop compared with repeated measures of testosterone in laboratory males. Horm. Behav. 5, 251-259 (1974).

Gartlan, J.S Structure and function in primate society. Folia. primatol. 8, 89-120 (1968)

Guhl, A.M. Gonadal hormones and social behaviour in infrahuman vertebrates, in Sex and Internal Secretions, Vol. II, W.C. Young, ed. Williams and Wilkins, Baltimore (1961), pp. 1240-1267.

Heller, C.G. and Nelson, W.O. Hyalinisation of seminiferous tubules and clumping of Leydig cells. Notes on treatement of the clinical syndrome with testosterone propionate, methyl testosterone and testosterone pellets. J. Clin. Endocrinol. 5, 27-33 (1945).

Kaufman, J.H. A three year study of mating benaviour in a free-ranging band of rhesus monkeys. Ecology 46, 500-512 (1965).

Keverne, E.B. Sexual and aggressive behaviour in social group of talapoin monkeys, in Sex, Hormones and Behaviour, Ciba Foundation Symposium 62. Excerpta Medica, Amsterdam (1979), pp. 271-286.

Keverne, E.B., Leonard, R.A., Scruton, D.M., and Young, S.K. Visual monitoring in social groups of talapoin monkeys (Miopithecus talapoin). Anim. Behav. 26, 933-944 (1978a).

Keverne, E.B., Meller, R.E., and Martinez-Arias. Dominance, aggression and sexual behaviour in social group of talapoin monkeys. Rec. Adv. Primatol. 1, 533-548 (1978b).

Kummer, H. Social organization of hamadryas baboons. Bibl. Primat., No. 6. Karger, Basel (1968).

Macrides, F., Bartke, A., and Dalterio, S. Strange females increase plasma testosterone levels in mice. Science 189, 1104-1106 (1975).

Mason, J.W. Organisation of psychoendocrine mechanisms. Psychosom. Med. 30 Suppl., 565-808 (1968).

Mason, J.W. and Brady, J.V. The sensitivity of psychoendocrine systems to social and physical environment, in Psychobiological Approaches to Social Behaviour, P.J. Leiderman and D. Shapiro, eds. Tavistock Publications (1965), pp. 4-23.

Matthews, L.H. Visual stimulation and ovulation in pigeons. Proc. Roy. Soc. B. 126, 557-560 (1938).

Michael, R.P. and Wilson, M. Effects of castration and hormone replacement in fully adult male rhesus monkeys (Macaca mulatta). Endocrinology 95, 150-159 (1974).

Nie, N.H., Hull, C.H., Jenkins, J.G., Steinbrenner, K., and Bent, D.H. SPSS: Statistical Package for the Social Sciences (2nd ed.). McGraw-Hill, New York (1975).

Phoenix, C.H., Slob, A.K., and Goy, R.W. Effects of castration and replacement therapy on the sexual behaviour of adult male rhesuses. J. Comp. Physiol. Psychol. 84, 472-481 (1973).

Rose, R.M., Bernstein, I.S., and Gordon, T.P. Consequences of social conflict on plasma testosterone levels in rhesus monkeys. Psychosom. Med. 37, 50-61 (1975).

Rose, R.M., Bernstein, I.S., Gordon, T.P., and Catlin, S.F. Androgen and aggression: a review and recent findings in primates, in Primate Aggression, Territoriality and Xenophobia, R.L. Holloway, ed. Academic Press, New York (1974), pp. 275-304.

Rose, R.M., Gordon, T.P., and Bernstein, I.S. Plasma testosterone levels in the male rhesus: influences of sexual and social stimuli. Science 178, 643-645 (1972).

Rose, R.M., Holaday, J.W., and Bernstein, I.S. Plasma testosterone, dominance rank and aggressive behaviour in male rhesus monkeys. Nature 231, 366-368 (1971).

Rowell, T.E. A quantitative comparison of the behaviour of a wild and a caged baboon group. Anim. Behav. 15, 499-509 (1967).

Rowell, T.E. Social organisation of wild talapoin monkeys. Am. J. Phys. Anthrop. 38, 593-598 (1973).

Rowell, T.E. and Dixson, A.F. Changes in social organisation during the breeding season of wild talapoin monkeys. J. Reprod. Fert. 43, 419-434 (1975).

Siegel, S. Nonparametric Statistics for the Behavioural Sciences. McGraw-Hill, Kagakusha Ltd., Tokyo (1956).

Steinberger, E. Hormonal control of mammalian spermatogenesis. Physiol. Rev. 51, 1-22 (1971).

Sugiyama, Y. Characteristics of the social life of bonnet macaques (Macaca radiata). Primates 12, 247-266 (1971).

Wilson, M., Plant, T.M., and Michael, R.P. Androgens and the sexual behaviour of male rhesus monkeys. J. Endocrinol. 52, ii (1972).

Wolfheim, J.H. and Rowell, R.E. Communication among captive talapoin monkeys (Miopithecus talapoin). Folia primatol. 18, 224-255 (1972).

Chapter 3
Studies in Adaptability: Experiential, Environmental, and Pharmacological Influences

E.N. Sassenrath

INTRODUCTION

Captive primates, living under a variety of colony environmental conditions, constitute effective models for the study of stress responses, particularly biobehavioral modes which determine whether the individual effectively adapts or succumbs to psychosocial environmental stress. In colonies which are self-perpetuating through effective breeding programs, it is also possible to carry out both longitudinal life-span studies and intergenerational observations to characterize correlates of effective adaptability under controlled colony social environments.

In such colonies, experience has shown that although infectious pathogens and environmental toxicants can be periodic factors in increased morbidity and mortality, a more pervasive correlate of poor health and reproduction is simply acute or ongoing psychosocial "stress." However, in all colonies, some animals survive better than others. The characterization of developmental or experiential factors which contribute to individual differences in effective adaptation has been the focus of several studies to be reported here. Also included will be observations of biological and behavioral indices which reflect the uniqueness of the individual's adaptation to its established group social environment: namely, individual differences among cagemates in patterns of stress hormone excretion levels and in behavioral responses to the drug delta-9-THC, the principle psychoactive component of marihuana.

TABLE I

LIFE SPAN CAGING ENVIRONMENTS OF RHESUS COLONY

Developmental Stage	Age Range	INDOOR CAGING		OUTDOOR CAGING	
		Single	Social Groups		
		Indiv. Cages N = 1 (+I)*	Wean. Groups N = 3 - 6	Crib Cages N = 6 - 20	Field Cages N > 40
INFANT:					
Prenatal	(in utero)	X		X	X
Post natal/ Preweaning	Birth - 6 mo.	X (1)	(1)	X	X
Postweaning	6-12 mo.		X		
JUVENILE:	9 mo.-2 yrs.			X (3)(2)	X
SUBADULT:	2-3 yrs.			X	X
ADULT: Females	> 3 yrs.	X	(4)	X	X
Males	> 4 yrs.	X	(4)	X	X

* Single adult or mother with infant (I)

N = Number animals per cage environment

MATERIALS AND METHODS

All animals studied were rhesus monkeys (<u>Macaca mulatta</u>) in the breeding program of the California Primate Research Center. Unless otherwise indicated they were colony born. All were housed in one of the three types of living conditions: indoor individual cages (IC), outdoor small group crib-type cages (OC), and half-acre large group or troop field cages (FC). Table one shows pathways of possible changes in caging environments during the lifespan of a colony-born monkey. For all animals, food (Purina Labchow) was given twice daily. Indoor caged animals were in a 12/12 hour light/dark schedule.

EXPERIENTIAL DETERMINANTS OF SOCIAL ADAPTABILITY IN INFANT, JUVENILE, AND ADULT RHESUS MONKEYS

<u>Weaning and Infant Peer Grouping</u>: For a colony-born infant, weaning or separation from

mother represents the first major adjustment to social environmental change. When infants are removed from the mother to facilitate breeding and are immediately grouped with other infants, they sustain not only a separation stress but the stress of an unfamiliar physical environment as well as unfamilar peer weanling cagemates. In our colony, when infants of six months of age were taken from individually caged mothers or from mothers caged in single male breeding cages and were immediately caged with other weanlings in indoor small group cages (see (1) in Table 1), there was a high incidence of morbidity and death within the first few weeks after weaning. When the cortisol stress response to this weaning/peer-grouping experience was monitored in these infants during the first week, a wide range of individual differences in the degree of duration of plasma cortisol elevation was found. Behavioral observations of these weanling groups over the subsequent months after group formation also indicated that those infants showing the highest cortisol response tended to be the more subordinate members of their peer groups.

In some groups composed of infants from outdoor group-caged mothers, cortisol response to weaning was very low, suggesting that prior social experience among adults had a modifying effect. In analysis of data from 27 such weanling/peer groupings, it was found that over 70 percent of infants showing higher than median cortisol response to weaning (>44 ng%) came from individually caged mothers, while 80 percent of those with lower than median cortisol reponses had been with group-caged mothers ($P<.025$) (Sassenrath, et al., 1976a). Behavioral differences between weanling groups from singly-caged mothers and group-caged mothers was also marked. Responses to an observer approach to the weanling cages were characterized by piloerection and huff-threatening from the outdoor infants, but by huddling, screeching, and grimacing by the indoor mother-reared infants.

Since the higher cortisol stress response to weaning was shown primarily by mother-caged infants --and since this high cortisol response to weaning was also shown to be a good predictor of subsequent

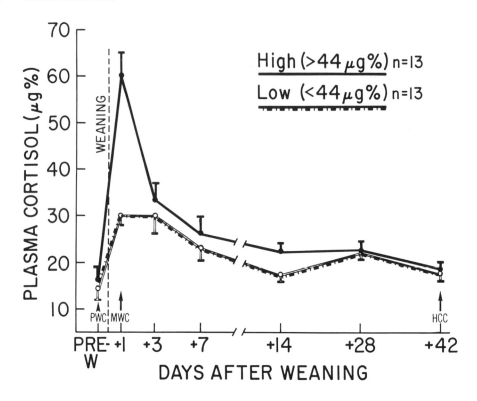

Figure 1. Mean post-weaning cortisol levels of high and low responders, based on 24-hour plasma cortisol response to separation from mother and simultaneous peer grouping. PWC = pre-weaning (with mother) plasma cortisol; MWC = maximum cortisol response to weaning/peer grouping; and HCC = home-cage cortisol six weeks after peer grouping.

health problems (Sassenrath, et al., 1976b) -- more infants from individually caged mothers were studied and their development monitored over a longer post-weaning period. It was found that infants showed the highest plasma cortisol levels one day after weaning. These levels were significantly reduced by the third day after weaning and peer-grouping and were stable at a lower constant level after one month. During this entire post-weaning period, however, those infants which were initially "high cortisol responders" continued to show significantly higher cortisol levels than the initial "low responders," as shown in Figure 1. For both high and low responders, their final stable mean group cortisol levels at six weeks (HCC) were still higher than their mean preweaning levels in caging with mothers (PWC).

In an attempt to modify the intense stress responses to weaning of the individually mother-caged infants -- and alleviate the social and health consequences -- one group of infants was weaned at four and one-half months of age by removal of the mother from the familiar home cage, and were peer grouped later at six months of age. In the group which remained in the familiar home cage environment, the cortisol response to separation from mother was much slower, so that plasma cortisol levels were higher at one week than at one day after separation. Further, the cortisol response to subsequent peer grouping nine weeks later was greatly reduced. This effect was particularaly striking when the plasma cortisol levels of these infants one day after peer grouping was compared to that of their like-aged cagemates undergoing peer grouping the day of weaning from mother, as shown in Figure 2. The behavioral correlates were also striking. Although all infants were undergoing their first socialization experience away from mother, those with lower cortisol elevations, who had adapted previoulsy to the stress of separation from mother, and to living without a "security figure", assumed the highest ranks (i.e., #1 and #2) within the three peer groups formed, while the high cortisol infants, showing predominantly submissive behaviors, assumed the most subordinate ranks. Although cortisol levels stabilized at much lower levels for all infants after one month in peer groups, the initial dominance ranking persisted (Sassenrath, 1977).

In other longitudinal observations of our colony-born rhesus monkeys, we have observed that the relative dominant/subordinate status assumed in a prior grouping tends to persist as animals mature and are moved to larger and/or older peer groups. This suggests a persistent influence of early experience in shaping biobehavioral adaptive patterns, particularly in the absence of any subsequent or intervening experiences to alter individual perceptions of new social challenges.

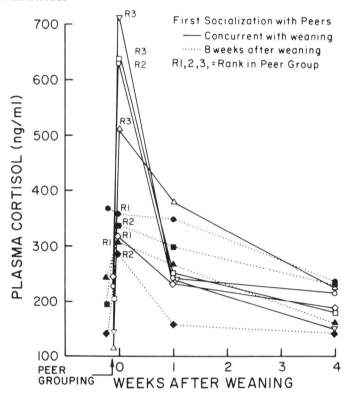

Figure 2. Cortisol stress response to first post-weaning socialization experience with peers for nine infants placed into three three-membered peer groups. Solid symbols represent infants weaned from mother none weeks prior to peer grouping. Open symbols represent infants peer-grouped the day of separation from mother. R1, R2, and R3 refer to dominance rank assumed by each infant in its peer group.

Juvenile Troop Formation from Three Age-Groups: The relevance of prior experience as a critical determinant of effective adaptation has also been confirmed in juvenile rhesus monkeys in a study of the formation of a larger juvenile field-caged group through combining three different-aged subgroups. The proposed field-caged juvenile "troop" was to be composed of 25 juveniles one year of age, 22 juveniles 16 months of age, and 12 juveniles aged two and one-half years. Both older groups were drawn from peer-groups of 24 to 27 animals housed in outdoor crib-type caging, while the youngest age-group was to be moved from their post-weaning indoor peer group caging of three to six animals each (see (2) and (3) in Table I).

To optimize the ability of younger age groups to adapt to older, larger cagemates in a strange outdoor environment, the subgroups were introduced to the large field cage sequentially, with the youngest animals first. Further, to enable the youngest group to adapt to physical aspects of the new environment prior to introduction of unfamiliar older cagemates, this group was exposed to the field cage environment intermittently during the first week; i.e., for six-hour intervals on three successive days, followed by overnight and weekend "recovery" periods in their familiar indoor home cages. They then lived full time in the field cage for one week to adjust to nighttime temperature changes before the introduction of the second age group juveniles.

Monitoring plasma cortisol levels during the first week at times of intermittent introduction and removal from the field cage gave a clear demonstration of elevation of this stress response after each six-hour period in the new environment followed by reduction of this stress reponse to lower basal levels after each "recovery" period in familiar indoor home cage environments. As shown in Figure 3, during the first week after each succeeding day in the field cage, cortisol elevation for the group declined until, after the fifth day in the field cage, the group mean cortisol level was not statistically significantly different from the pre-grouping home cage level.

Behavioral correlates of this process of adaptation to the new environment were consistent with the cortisol data. Namely, on the initial days when cortisol was high, all animals remained in sheltered corner areas of the field cage. Social interaction was limited to huddling together and low intensity aggressive interaction (threats and yields) initiated by dominants to unfamiliar cagemates. As cortisol levels decreased during subsequent days, social interaction changed to active play, with general utilization of all peripheral cage structures as well as structures in the exposed

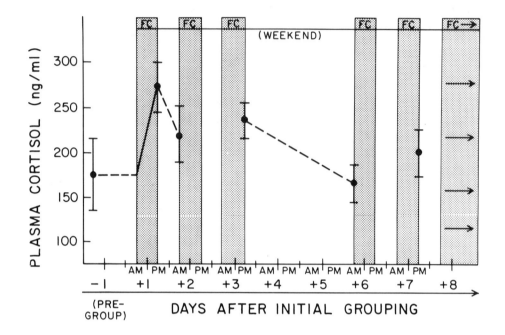

Figure 3. Plasma cortisol response to intermittent six-hour session in field cage (FC) environment alternating with overnight and weekend periods in familiar indoor small-group cages. Points are mean values for eight of 25 year-old juveniles, selected to be equally representative of males and females and dominant and subordinate group members.

central grassy areas. Significantly, group responses to disturbances outside the cage (low flying planes, motorized grass cutting and approach of observers) also changed from stationary huddling-cringing to active running escape.

As subsequent older peer groups were introduced at weekly intervals thereafter, this younger group was able to utilize effective active escape routes in response to attempts to aggress on them, surviving with minor trauma (wounding) and no delayed morbidity.

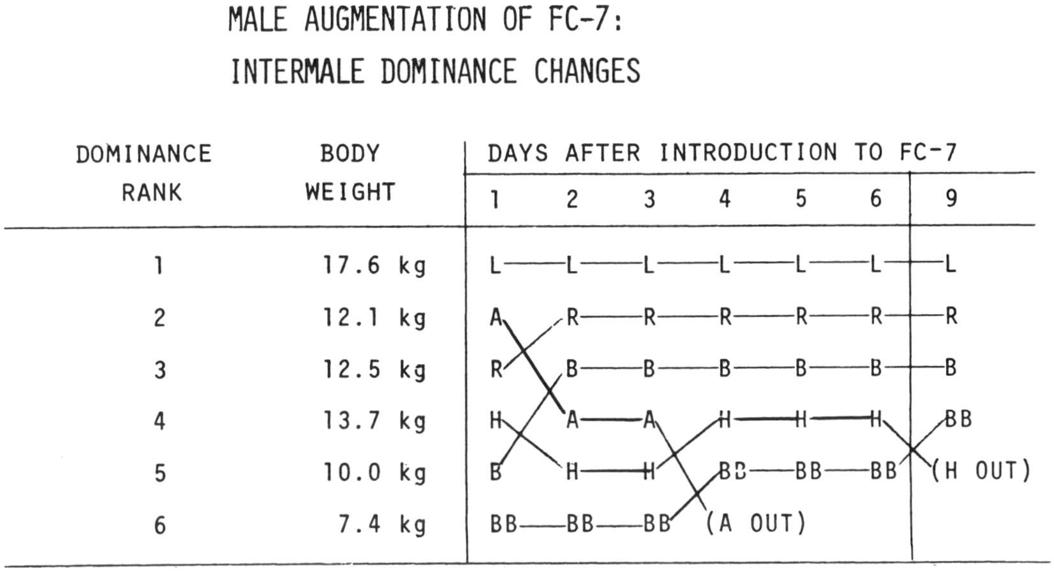

MALE AUGMENTATION OF FC-7: INTERMALE DOMINANCE CHANGES

DOMINANCE RANK	BODY WEIGHT	DAYS AFTER INTRODUCTION TO FC-7						
		1	2	3	4	5	6	9
1	17.6 kg	L	L	L	L	L	L	L
2	12.1 kg	A	R	R	R	R	R	R
3	12.5 kg	R	B	B	B	B	B	B
4	13.7 kg	H	A	A	H	H	H	BB
5	10.0 kg	B	H	H	BB	BB	BB	(H OUT)
6	7.4 kg	BB	BB	BB	(A OUT)			

Figure 4. Changes in dominance status for six adult male augmentees to field cage troop of feral females.

Although this observational study did not include a comparison group in which similar age-groups were grouped simultaneously, prior experience in transfer to outdoor caging would have predicted some incidence of both wounding and illness within the first week. Thus, again, the effectiveness of immediate prior experience in developing coping skills and modifying appraisal of coping resources, was a strong determinant of adaptability.

<u>Adult Troop Augmentation</u>: The element of prior relevant experience has also been shown to be critical in the adaptation of older adult males to introduction to an established troop of females in a field cage environment. In this instance, six adult breeder males were added simultaneously to an established group of 60 females (see (4), Table I). Of these six males, four survived to become the male

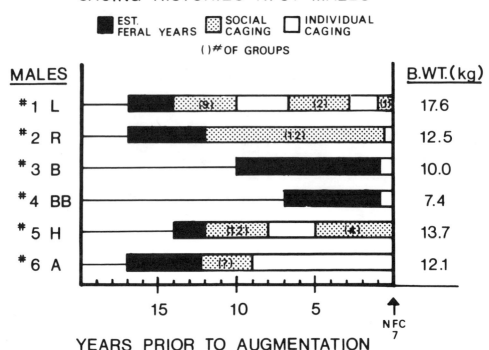

Figure 5. Caging histories of male breeder augmentees added to established field cage troop of 60 females. Numbers in parenthesis refer to number of harem groups placed with that male during the specified period.

breeder component of the troop. The day-to-day success of their initial social adaptation is shown in Figure 4.

Clearly, the initial inter-male dominance ranking on day one showed a high correlation with body size. This relative ranking persisted, except for males A and H who were severely wounded and were removed within ten days after introduction to the cage.

A correlation with effective survival in this situation was found in their past caging histories, as shown in Figure 5. All males were wild-born, but although A and H were not the smallest or youngest, they were clearly those with the least relevant complex social experience in their immediate past environments. A, who was first to go after the male group introduction, had had only individual pair-breeding social contact in the indoor breeding colony for nine years prior to this time. Similarly, H, who went out after one week, had had only limited socialization in four small single-male breeding groups during the five previous years. The other males were either recently from the wild, as the two younger males ranked #3 and #4, or had experienced more diverse environmental changes with more female groups in the years prior to the move to the field cage, as the older males ranked #1 and #2.

In these four males, the endocrine responses to their introduction to the field cage reflected their individual behavioral adaptation. As shown in Table II for all but the alpha male, the circulating plasma cortisol level one month after introduction to the field cage was higher than that before introduction. Further, after one month, the levels were elevated in inverse order to their relative dominance ranking, with the stress of day-to-day adaptation to the ongoing social environment most apparent in the most subordinate, smallest male. Similarly, the increase in cortisol response to a standard ACTH challenge given before and after one month in the field cage reflected the greater chronic adrenocortical activation of the subordinates in their new social environment (Sassenrath and Goo, 1978, Sassenrath, 1971).

During this same month, as shown in Table II, the testosterone response of these males to the sexual stimulation of female cagemates during the Fall breeding season was also suppressed. This was evident in comparison with testosterone elevations for the breeder males in an established breeding troop in a neighboring field cage. It is also of interest to note that this general stress-related suppression of testosterone levels was not apparent one year later.

This type of correlation among levels of activation of different endocrine systems and between selected social (subordinate) behaviors and endocrine systems had been repeatedly verified in our observation of rhesus cage-groups, whether three-to-four membered small indoor cage-groups or large 60 to 80-membered outdoor groups (troops).

TABLE II

PLASMA CORTISOL AND TESTOSTERONE LEVELS IN TROOP-CAGE MALES BY RANK:
NEW AUGMENTEES IN FC-7 vs. ESTABLISHED MALES IN FC-3

CAGE	MALES BY RANK	CORTISOL (ng/ml)		TESTOSTERONE (ng/ml)	
		BEFORE (9/77)	1 MO. AFTER (10/77)	BREEDING SEASON (10/77)	(10/78)
FC-7	#1	141	138	5.1	4.7
	#2	107	178	4.8	8.0
	#3	141	266	5.9	11.1
	#4	185	303	5.0	7.4
	MEAN			5.2	7.80
	S.D.			±0.48	±2.62
FC-3	#1			11.0	13.4
	#2			12.0	8.4
	#3			4.6	5.6
	#4			9.6	5.6
	MEAN			9.30	8.25
	S.D.			±3.28	±3.67
	*td			.47	0.03
	*p			.025	N.S.

*t value for the difference between means and P values taken from table of microstatistics for small samples (Dixon & Massey, 1951, p. 242).

Specifically, with appropriately standardized observational procedures, one can show a consistent correlation between frequencies of relevant subordinate behaviors and indicates of adrenocortical activation, although the correlation of either with rank, per se, may be less than 1.0, especially if males and females are included in one ranking for the group.

NEUROENDOCRINE PROFILES REFLECTING ADAPTIVE MODES AMONG CAGEMATES

Although it is possible to show a degree of correlation among group members for certain selected indices of social adaptation -- such as subordinate behaviors, adrenocortical activation, and male testosterone levels -- the overall patterning of other neuroendocrine stress indices becomes more complex. This can be demonstrated by data from four small cage-groups for excretion levels of epinephrine, norepinephrine and cortisol during 24 hours following removal from the group

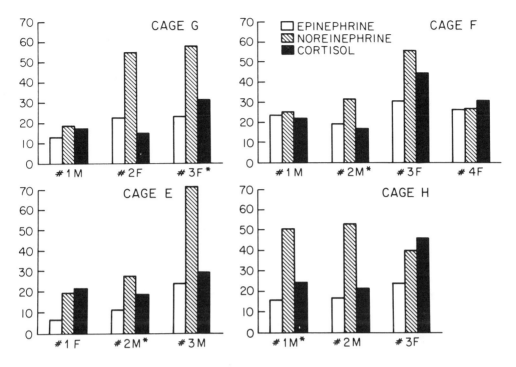

Figure 6. Excretion patterns of stress hormones for individual members of four cage groups. Numbers refer to dominance rank; M = male; F = female.

cage environment. As shown in Figure 6, although the most dominant animals in each group tend to be lowest for each hormone, and the more subordinate ranks tend to be higher, there is not a consistent trend among hormone levels within or between individuals. However, in the overall picture, each endocrine pattern can be viewed as an appropriate correlate of the behavioral profile of that individual, reflecting both activity and arousal levels as well as the emotional component or "intensity" of that individual's adaptation to its social environment (Sassenrath, 1969). Although this pattern of adaptation is known to include endocrine systems (Mason, 1974), each changing to accommodate the demands on that organism and optimize his survival, two response systems appear to constitute unique indicators of the organism's appraisal and "coping stance" to his immediate and/or ongoing psychosocial situation. These have been called traditionally the "fight or flight" response to stress, characterized by sympathetic activation, and the hypothalamic-pituitary-adrencortical axis response to stress, characterized by ACTH and cortisol or corticosterone secretion.

The relative significance of these two pathways in psychosocial adaptation can be expressed as in Figure 7. Here the perceived stimulus or situation is appraised on the basis of past experience to activate a coping mode which can involve both response pathways, but favoring one over the other depending on the response resources available to the organism. Where appraisal results in "fight or flight" response with active coping behaviors, there is concurrent sympathetic activation, and endocrine changes appropriate for arousal and energy mobilization. When appraisal indicates no immediate control over the stressor -- that nothing the individual can do will have any effect -- the "conservation-withdrawal" response predominates, with submissive avoidance behaviors and neuroendocrine changes appropriate for conservation and restoration of energy resources. These would include pituitary-adrenocortical activation and -- in the absence of sympathetic stimulation -- predominantly parasympathetic activity.

So, in psychosocial adaptation, each individual in a given environment assumes a unique patterning of internal responses depending on his role in the group and his adaptation to group interaction. It also follows that there would be uniquely selective activation of pathways in the brain that integrate that individual's appraisal and his behavioral and physiological responses to his environment (Henry, 1977).

EFFECTS OF DELTA-9-THC ON SOCIAL ADAPTATION TO GROUP CAGE ENVIRONMENTS

The effects of delta-9-THC, the principal psychoactive component of marihuana, on the social

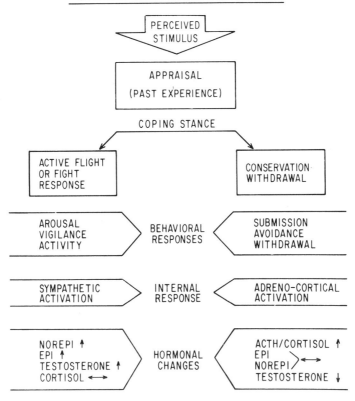

Figure 7. Behavioral and neuroendocrine correlates of "Fight and Flight" or "Conservation-Withdrawal" coping stances in psychosocial stress responses (adapted from J.P. Henry, 1977).

behavior of group-living monkeys is characterized by a high degree of inter-individual variability and a paradoxical reversal of effects between acute or short-term and long-term chronic exposure to the drug. One consistent aspect of this drug lies in its ability to alter characteristic adaptive social behavior patterns of individual members of established social groups. From our own observations we have concluded that the extreme individual differences in response to the drug are a function of at least two prime variables: 1) predetermined individual differences in social adaptability interacting with the immediate social setting, and 2) the level of CNS-mediated "tolerance" of the individual to THC, determined by the extent and duration of prior exposure to the drug.

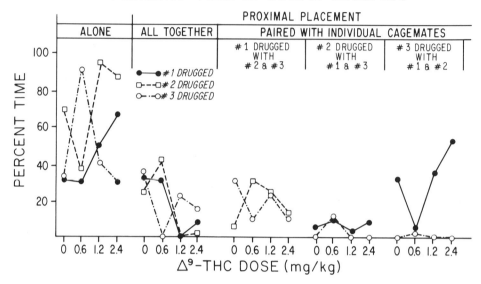

Figure 8. Changes in social spacing for each member of a three-membered cage group at increasing dose levels of delta-9-THC. Proximal placement indicates body positioning within touching distance of another cagemate.

Social spacing between or among individual members of a social group is one index of inter-individual adaptation to the stress of forced proximity in caged environments. As shown in Figure 8, the individual members of one three-membered cage group can differ markedly in the percent of time they spend sitting alone, or proximal (within touching ditance) to one or both cagemates. Figure 8 also shows how the patterns of social distances change from each group member after exposure to increasing doses of THC. While the #1 and #2 ranking males tend to spend more time "alone" or out of contact with cagemates at higher doses of THC, the #3 female shows this effect only at the lowest THC dose; and while the #1 male shows reduced contact with #2 male at higher THC doses, the #3

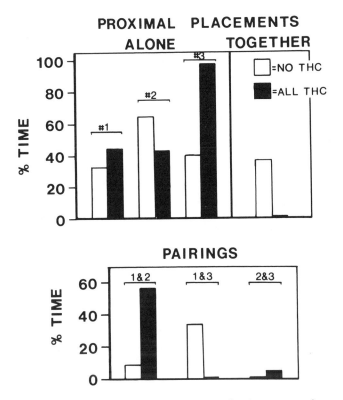

Figure 9. Differences in social spacing for a three-membered cage-group between the normal undrugged condition and during the five hours after all members received delta-9-THC orally at 2.4 mg/kg.

female shows increasing contact with the #1 male at higher drug levels. Clearly, this type of non-linear dose-response relationship cannot be explained on the basis of a simple CNS depressant effect characteristic of sedative-hypnotic type drugs, although initial THC intoxication is characterized by reduced activity in most subjects (Chapman, et al., 1978).

Rather, the effects of the drug on social interaction of group-living monkeys can be better characterized as a change in the perception and/or processing of social stimuli in the drugged group member resulting in changes in adaptive responses in group interaction. This is further demonstrated by the changes in social spacing when all members of this group are drugged simultaneously at the highest dose level of THC, as shown in Figure 9. Under these conditions, the #3 female withdraws entirely, while the #1 and #2 males markedly increase their mutual proximal placements. Here, social preferences or adaptive avoidances operating under normal conditions are completely altered by the drug.

In intra-group agonistic interactions, not all group members respond in the same way, nor do all animals of similar rank in different cage groups respond in the same way. However, as shown in Table III, after acute THC administration at 2.4 mg/kg, most dominant group members (ranked #1 and #2) in the five cagegroups studied tended to show fewer aggressive behaviors, both in spontaneous non-stimulated group interaction and in competition for preferred food. However, some of these subjects showed a paradoxical increase in the more active hit/bite/chase aggressive behaviors after acute THC exposure.

This latter type on intensification of agonistic responses becomes the predominant response to the drug as daily drugging is continued on a long-term basis. As shown in Table III, when daily drugging of #1 and #2 ranking single members of four cagegroups was continued through the course of one year, all showed an increase in the frequency of hit/bite/chase aggressive behaviors and/or more intense attack episodes. In contrast, all undrugged animals of comparable rank in the same cagegroups showed decreases or no change in frequencies of active aggressive behaviors.

THC EFFECTS ON SOCIAL BEHAVIOR AS A FUNCTION OF DRUG TOLERANCE

The data presented indicate that there is a major qualitative change in THC effects on social behavior with continued long-term drug exposure: namely, a transition from the transient predominant tranquilizing effects of initial intoxication to a pervasive irritable over-responsiveness which is intensified under conditions of social stress. Our initial occasional observations of increased

TABLE III

INDIVIDUAL DIFFERENCES IN DELTA-9-THC INDUCED CHANGES IN FREQUENCIES
OF AGGRESSIVE BEHAVIORS: ACUTE VS. LONG-TERM CHRONIC DRUG
EXPOSURE VS. UNDRUGGED CAGEMATES

DRUG STATUS	AGONISTIC INTERACTION	DIRECTIONAL CHANGE IN FREQUENCY OF AGGRESSIVE BEHAVIORS		
		DECREASE	NO CHANGE	INCREASE
ACUTE THC*	SPONTANEOUS			
	Threat	5/9**	4/9	
	Hit/Bite/Chase	4/9	2/9	3/9
	Attack			
	COMPETITION			
	Threat	7/9	3/9	
	Hit/Bite/Chase	6/9	2/9	1/9
LONG-TERM CHRONIC THC*	SPONTANEOUS			
	Threat			
	Hit/Bite/Chase	1/4	1/4	2/4
	Attack			4/4
	COMPETITION			
	Threat			
	Hit/Bite/Chase			4/4
LONG-TERM UNDRUGGED CAGEMATES	SPONTANEOUS			
	Hit/Bite/Chase	5/7	2/7	
	COMPETITION			
	Hit/Bite/Chase	2/7	5/7	

*Delta-9-THC oral dosage at 2.4 mg/kg/day

**Fractions refer to the number of drugged or undrugged animals ranked #1 and #2 of all groups studied which showed the indicated change in frequency of aggressive behaviors. <u>Acute</u> data summarizes behavioral interactions observed at 1, 3, and 5 hours after a single dose of THC. <u>Long-term chronic</u> data summarizes behavioral data taken continuously over a one year period after initiation of daily THC drugging, with each behavioral observation session at 20 hours after the prior days drugging.

aggression in some newly drugged animals, suggest that a selective CNS-mediated tolerance builds up over time to the tranquilizing effects of THC, unmasking the stimulant or aggression-enhancing properties of the drug. This tolerance is dose-dependent, and can be overcome by exposure to higher levels of the drug.

The observations reported here and elsewhere (Sassenrath and Chapman, 1975, 1976; Chapman, et al., 1979) demonstrate that when THC dosage exceeds the level of THC tolerance in the brain (as in acute or infrequent exposure to the drug) the well-known tranquilizing-sedating manifestations of THC intoxication occur. This results in reduction in competitive behavior, active play, and assertion of dominance. However, with continued chronic exposure to THC, a dose-dependent tolerance develops in the brain which results in an irritability and over-responsiveness, even in the same familiar social environment. Under these conditions of stress, this can erupt as violent aggressive episodes.

In the study of effects of long-term drugging of selected group members on group social interaction, the frequencies of observed submissive responses of undrugged cagemates increased as the frequencies of aggresive behaviors of drugged cagemates increased. These responses of undrugged cagemates ultimately proved to be the more sensitive index of behavioral change in the drugged animals (Chapman, et al., 1979). By varying the hour of daily drugging and holding observation schedules constant, it was possible to demonstrate, through frequencies of these submissive responses, that the behavioral changes were not related to withdrawal effects preceding each daily drugging. Rather, the overall changes in social interaction concurrent with long--term daily exposure of single cagemates to THC, appeared to result from an increased irritability and responsivity to social stresses which altered adaptive behaviors in ongoing social interaction.

A demonstration of stress enhancement of aggressive behavior in the chronically drugged animals occurred in subsequent studies of reproductive function in THC treated females. Two single-male breeding groups were formed, each including two long-term THC-treated adult females and three non-drugged females. In the first group, the smaller of the THC-treated females was killed overnight within the first 24 hours of grouping. The second cage was more carefully monitored and such high aggression by the THC-treated females was documented that they were removed overnight and for longer intervals to permit gradual adaptation among unfamiliar cagemates. During these periods of introduction and removal of the drugged females, blood samples were taken from all females to determine stress-related changes in circulation cortisol. The data, as shown in Figure 10, confirmed the differential between chronically THC-drugged females and undrugged cagemates. While all females showed elevated cortisol response to the new group environment, the mean levels of the undrugged females were consistently and significantly higher when ever the THC-females were in the cage, in apparent response to their hyper-responsive, aggressive interaction behaviors which ultimately necessitated their complete removal from the group.

In both conditions -- that of THC-tolerant or non-tolerant brain -- THC appears to interfere with perception and/or the processing of social stimuli, resulting in altered behavioral responses to ongoing intragroup social stimuli. Where a given dose of THC may produce increased passive affiliation or withdrawal in naive subjects, the same dose may produce aggressive outbursts in drug tolerant subjects in the same social environment. This change in the predictability of response is especially apparent to non-drugged cagemates and is manifested by them as increased submission and avoidance of drugged group members.

Because of the long half-life of the highly lipid soluble THC in lipid-rich tissues such as the brain, it is not surprising that such a change occurs with chronic drug intake, due to the high retention of the drug in such tissues. There are well documented observations of the drug's interference with cognitive and perceptual processing -- and there are accumulating clinical observations of

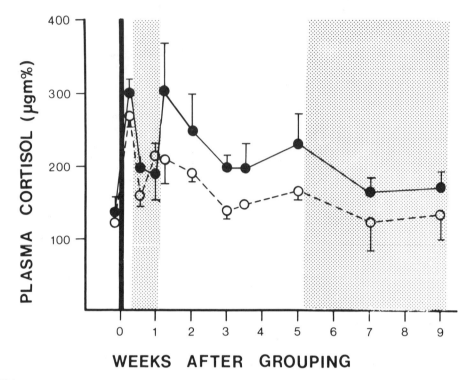

Figure 10. Stress-elevation of mean plasma cortisol levels of undrugged females (N=3) and long-term THC-drugged females (N=2) after new group formation. Solid circles, drugged and open circles undrugged. Shaded areas represent periods when drugged females were removed from the cage each night for the 18-hour period before blood sampling at 9:00 a.m.

the persistence of decrements in cognitive processing after long-term exposure to marihuana in man. It is at this level of alteration of processing environmental stimuli that the effect of the drug on social behavior appears to be mediated as well, thereby altering the individual's application of his acquired adaptive behaviors. To the extent that these are extremely variable among individuals, it follows that specific behavioral manifestations of THC effects will also be variable, depending on the individual's past experience in interaction with his current social situation -- as well as the pharmacological variable of acquired tolerance.

CONCLUSIONS

These studies have demonstrated that effective biobehavioral adaptation to the psychosocial stress of new living environments in colony-living rhesus monkeys is a direct function of relevant prior experience. In the wild, such relevant experience and learning takes place gradually in the course of normal development and maturation. For captive or colony-living animals, opportunities are limited for developing coping resources to respond effectively to novel or changing environments. However, such opportunities for learning can be simulated by appropriate manipulation of animals through relevant experience.

Once adjustment has been made to living in an established cage group, the patterning of neuroendocrine and behavioral responses to ongoing group interaction can vary dramatically among group members, reflecting the unique role of each in the group social structure. The striking reciprocal relationship between activation of the pituitary-adrenal axis characteristic of subordinate animals and suppression of testosterone response to sociosexual stimuli has been demonstrated here with rhesus monkey and confirmed by Keverne's studies with the Talapoin monkey (Keverne, 1979; and this volume). Keverne's work further demonstrates that such inter-individual differences related to social rank are mediated through alterations of neuroendocrine mechanisms resulting in rank-related differences in CNS responsivity to hormonal stimuli and to pharmacological intervention which selectively alters neuronal function in the CNS.

Our own observations on the behavioral effects of delta-9-THC, the principal psychoactive component of marihuana, demonstrate a pharmacological action which is also greatly influenced by the psychosocial adaptation of each animal to its group environment. In this case, however, the action of this drug can best be described as an alteration of normal processing of perceived social stimuli resulting in marked deficits in the application of previously acquired adaptive behaviors.

A further striking characteristic of this drug is its ability to cause contrasting changes in social behavior during the course of continued drug exposure. These slowly developing changes -- from predominant tranquilization to irritability -- are mediated presumably through changes in CNS neuronal function in adaptation to persistent levels of the drug. With long-term exposure to the drug, the impairment of effective application of learned behavioral responses is most apparent in situations of high psychosocial stress, where adaptive behaviors are most critical. Under such conditions, where long-term THC intake interacts with a stressful social environment -- non-adaptive irritable aggressive behaviors of drugged cagemates can elicit both avoidance/escape behavior and elevated stress responses in undrugged cagemates and the disruption of group social structure.

ACKNOWLEDGEMENTS

I wish to acknowledge the major contribution of my colleagues Gail P. Goo, Thomas I. Madley, and Dr. Mari S. Golub, to the studies reported here. Major responsibility for field cage observations and the conduct of weanling studies was carried by students George Collins, Ray Avels, and Jim Wallace. The cooperation of Roy Henrickson, D.V.M., Walter S. Tyler and Andrew G. Hendrickx, administrative officers of the California Primate Research Center, in facilitating and supporting these studies is gratefully acknowledged. This work was supported by USPHS grants RR00169 and DA00135.

REFERENCES

Chapman, L.F., Sassenrath, E.N., and Goo, G.P. Social behavior of rhesus monkeys chronically exposed to moderate amounts of delta-9-tetrahydrocannabinol, in Advances in the Biosciences, Vol. 22 and 23: Marinuana: Biological Effects, G.G. Nahas and W.D.M. Paton, eds. Pergamon Press, New York (1979).

Dixon, W.J. and Massey, F.J. (ed.) *Introduction to Statistical Analysis*. (1951).

Henry, J.P. and Stephens, P.M. (eds.) *Stress, Health and the Social Environment*. Springer-Verlag, New York (1977).

Keverne, E.B. Sexual and aggressive behaviour in social group of talapoin monkeys, in *Sex, Hormones and Behaviour*, Ciba Foundation Symposium 62. Excerpta Medica, Amsterdam (1979), pp. 271-286.

Mason, J.W. Specificity in the organization of neuroendocrine response profiles, in *Frontiers in Neurology and Neuroscience Research*, P. Seeman and G.M. Brown, eds. University of Toronto Press, Toronto (1974).

Sassenrath, E.N. Neuroendocrine adaptation to the stress of group caging in M. mulatta. *The Physiologist* 12, 348 (1969).

Sassenrath, E.N. Increased adrenal responsiveness related to social stress in rhesus monkeys. *Hormones and Behavior* 1, 283-298 (1970).

Sassenrath, E.N., Henrickson, R.V., Goo, G.P., Golub, M.S., and Madley, T.I. Cortisol response to weaning stress and subsequent health record of colony born rhesus juveniles. *Proceedings of the 27th Annual Session of the American Association of Laboratory Animal Science*, Houston, Texas, November (1976a).

Sassenrath, E.N., Goo, G.P., Golub, M.S., Hendrickx, A.G., and Russell, J.L. Antecedents and consequences of endocrine stress response to weaning in colony-born rhesus infants. *Proceedings of the 6th Annual Meeting of the Society for Neurosciences*, Toronto, November (1976b).

Sassenrath, E.N. Weaning strategies and behavioral sequelae in colony born rhesus infants. *Proceedings of the Inagaural Meeting of the American Society of Primatologists*, Seattle, Washington (1977).

Sassenrath, E.N. and Goo, G.P. Augmentation of established domestic breeding groups of rhesus macaques. *Proceedings of the 29th Annual Session of the American Association of Laboratory Animal Science*, New York, September (1978).

Chapter 4
Social Status Related Differences in the Behavioral Effects of Drugs in Vervet Monkeys *(Cercopithecus Aethiops Sabaeus)*

M.J. Raleigh, G.L. Brammer, M.T. McGuire, A. Yuwiler, E. Geller, and C.K. Johnson

INTRODUCTION

This chapter reviews aspects of the effects of differences in social status on the behavioral responses to drugs affecting primarily serotonergic systems in adult male vervet monkeys (Cercopithecus aethiops sabaeus). We will first discuss some features of the social behavior of vervet monkeys which make them appropriate subjects for this type of study. Subsequently, to provide a rationale for the choice of drugs and behavioral parameters, we will review our previous pharmacological-behavioral studies as well as those on relationships between behavior and peripheral biochemical parameters which may reflect central serotonergic activity. Finally we will present data illustrating the interplay between social status, environmental test situations, and the behavioral responses to drugs in vervet monkeys.

VERVET SOCIAL BEHAVIOR

Three features of vervet monkey social behavior make these animals appropriate subjects for examining the effects of social status on the behavioral responses to drugs. First, in contrast to many other species of Old World monkeys, vervets tolerate captivity well. The effects of captivity on social behavior appear to be quantitative rather than qualitative (Fairbanks and Bird, unpublished). In large outdoor enclosures, captive vervet groups engage in behaviors that resemble those seen in free-ranging situations (Fairbanks et al., 1978; Rowell, 1971; Struhsaker, 1967). Although the rate of affiliative behaviors appears to be increased, the patterns of mother-infant interactions, caretaking of neonates, sex differences in play, and spatial relationships seem unaltered by captivity (Lancaster, 1971; McGuire et al., 1978; Raleigh et al., 1979).

Second, in free-ranging settings vervets often live in multimale groups. Troops with up to 65 animals containing as many as nine adult males have been reported (McGuire, 1974). Similarly, in captivity, we have maintained groups containing up to 35 animals, six of which were adult males. This tolerance of more than one adult male in captive groups contrasts with several other species of Old World monkeys including patas monkeys and most types of baboons.

A third feature of vervet monkeys is that one adult male, the dominant male, can be clearly distinguished from all other males in both captive and free-ranging groups (Fairbanks and McGuire, 1979; Struhsaker, 1967). This differentiation is consistent across a variety of measures including: (1) agonistic rank, (2) control male factor scores, and (3) access to limited resources; it is also apparent qualitatively. For example, dominant males differ from other males in their postures and in the attention given to them by other group members. In captivity when group membership is stable, status relationships may persist for up to five years (the longest period examined to date). While a dominant male can be easily recognized, the remaining nondominant adult males cannot be consistently rank-ordered in social status. Although nondominant males are clearly not behaviorally homogeneous, their ranking across behavioral measures is inconsistent. A nondominant male that scores high on the agonistic rank measure may have a low control male factor score. Thus, unlike adult males in some other species of Old World monkeys, adult male vervets seem to exhibit a binary rather than a linear dominance hierarchy. This binary sorting of adult male vervet monkeys into dominant males and other males has both biochemical and behavioral correlates.

Figure 1 shows that dominant males are distinguished by a difference in at least one peripheral biochemical parameter. This figure shows the mean ($\pm$ standard error) of the concentration of whole blood serotonin in adult male vervets. When adult males from 12 established captive groups, each containing at least three adult males, were examined, the 12 dominant males had significantly higher levels of whole blood serotonin than the 28 other males. In addition, in each of the 12 groups, the male with the highest whole blood serotonin concentration was dominant. When status relationships remain constant, whole blood serotonin does not vary greatly within an individual. In contrast, differences between individuals are large. Each of these 40 animals was sampled at least three separate times and the comparisons reported above are based on the mean value for the individual. Neither circadian nor seasonal rhythms have been observed. Because both dominant and nondominant males received identical diets and hoarding did not occur, the whole blood serotonin differences between dominant and other males do not appear to be due to differences in food consumption (McGuire et al., in press a).

Although the correlation between dominant male status and high whole blood serotonin concentration has been strikingly consistent, its significance is currently unknown. It is unclear whether this high value represents a trait of the individual animals or is a consequence of occupying the dominant male social position. Moreover, it is not known whether whole blood serotonin concentrations change when dominance relationships are manipulated as, for example, by adding or removing animals from the group. It is also not known whether current whole blood serotonin levels in individually housed subjects predict either their future serotonin levels or their dominance status in subsequently formed social groups. Furthermore, the mechanisms responsible for the high whole blood serotonin/dominant male association have yet to be investigated. There are no data on the differences between dominant and other males regarding the synthesis,

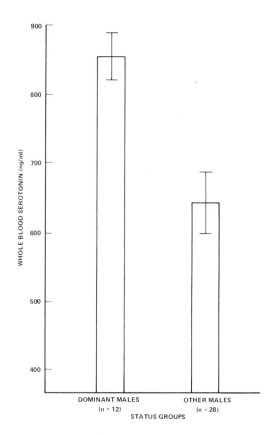

Figure 1. The mean (± standard error) concentration of whole blood serotonin (ng/ml) in 12 dominant males and 28 other males.

storage, or degradation of whole blood serotonin which may account for the observed status-related differences in whole blood serotonin level (McGuire et al., in press b). While the relationship between dominant status and whole blood serotonin seems clear, it is unlikely that whole blood serotonin is the only peripheral biochemical parameter which distinguishes dominant males from other males. Sassenrath (1980) and Keverne (1980) have reported that differences in social status in rhesus and talapoin monkeys may be related to differences in basal and challenged cortisol, prolactin, and testosterone concentrations. These measures have yet to be examined in vervet monkeys.

As shown in Figure 2, there are other behavioral consequences of the differentiation between dominant and nondominant males. This figure shows the rate of grooming behavior among non-treated

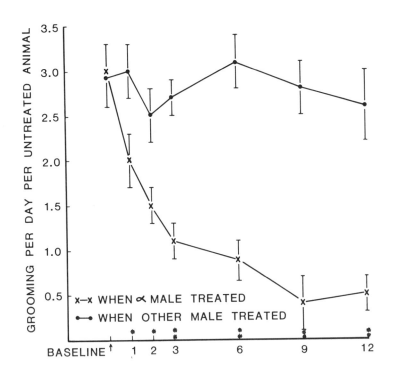

Figure 2. Alterations in grooming behavior in untreated group members when either a dominant or a nondominant male is treated with PCPA. The mean rates of grooming ($\pm$ standard error) per day per untreated animal are shown. *=$p<.05$, **=$p<.01$.

group members when one adult male group member is treated with p-chlorophenylalanine (PCPA). When vervet monkeys are treated with PCPA, they are more aggressive, are more solitary, locomote more, are more vigilant, groom less, approach less, eat less, and rest less. These changes are seen in both dominant and other males. However, as Figure 2 shows, the impact of these changes on the behavior of other group members varies with whether a dominant male or nondominant male is treated (McGuire et al., in press b). This figure summarizes data from six groups, each containing three adult males and three adult females. One male was treated at a given time with 80 mg/kg of PCPA daily while the other five animals received saline. When a dominant male was treated, the untreated animals were attentive to the alterations in his behavior and the frequency of their affiliative behaviors such as grooming declined. By the second day, their grooming had diminished to 50% of the pretreatment levels. This effect continued throughout the treatment period; there was no habituation to the alterations in the behavior of the dominant male. In contrast, when a nondominant male was treated, the grooming behavior of the other group members was unaffected. These data are compatible with Redmond's observation (1980) that following AMPT treatment, untreated group members exhibit striking alterations in behavior.

RATIONALE

The behavioral measures and pharmacological manipulations stem from our interest in assessing the relative contribution of serotonergic systems to the mediation of vervet social behavior. In this work we have been guided by two assumptions. First, we assumed that all behaviors are mediated by a variety of transmitter systems. The initiation, maintenance, and termination of any behavior results from the interplay among numerous transmitter systems; no behavior is the result of the action of a single transmitter system. Therefore, we have tried to determine the relative, rather than absolute, contributions of serotonergic mechanisms to behavior (Raleigh et al., 1980). The second assumption is that no drug is totally specific for a given transmitter system. The specificity ascribed to a drug is often inversely proportional to what is known about it. For example, PCPA treatment was initially regarded as a fairly specific means of reducing serotonin biosynthesis (Koe and Weismann, 1966). While PCPA is a noncompetitive inhibitor of tryptophan hydroxylase and does indeed impair serotonin biosynthesis, it also has a number of other properties. As a neutral amino acid, PCPA competes with other neutral amino acids for entry into the brain across the blood brain barrier via the neutral amino acid carrier (Oldendorf, 1971). Thus PCPA may alter the amount of tyrosine, phenylalanine, leucine, tryptophan, and other neutral amino acids reaching the brain and consequently may affect the activity of a variety of neurotransmitters in addition to serotonin. Further, during the initial 48 to 72 hours after administration, PCPA competitively inhibits tyrosine hydroxylase, an enzyme involved in catecholamine biosynthesis. PCPA also initially competitively inhibits tryptophan hydroxylase before becoming a non-competitive inhibitor (Jequier et al., 1967). Finally, PCPA acts peripherally as well as centrally. Thus, while PCPA does reduce serotonergic function, its activity is hardly specific to central serotonergic neurons. Because no drug has complete specificity, we have employed a battery of drugs directed largely at serotonergic systems and utilized the commonalities of their behavioral effects to make inferences about the relative serotonergic contribution to behavior (McGuire et al., in press b).

In our pharmacological behavioral studies, we employed drugs as a means of altering serotonergic mechanisms. To that end, for each drug we administered the lowest dose that consistently altered behavior on the basis of prior dose-response testing. This approach differs from that of Claus and Kling (1980), who investigated the effects of an abused substance on monkey behavior, in that the drugs per se were of secondary interest while our primary interest was in how they affected serotonergic mechanisms. Figure 3 illustrates the primary sites of action of the drugs used: tryptophan, 5-hydroxytryptophan, PCPA, and chlorgyline, alone or in combination. Tryptophan is a precursor to serotonin, and once it has crossed the blood brain barrier, it is 5-hydroxylated only in serotonergic neurons (Moir and Eccelston, 1968). The resulting 5-hydroxytryptophan can then be decarboxylated to serotonin (Yuwiler, 1973). Thus, tryptophan administration should enhance serotonergic function. We also gave PCPA, a noncompetitive inhibitor of tryptophan hydroxylase which thereby reduces serotonin biosynthesis (Koe and Weismann, 1966). We administered 5-HTP alone or in combination with PCPA. By itself, 5-HTP is a precursor of serotonin.

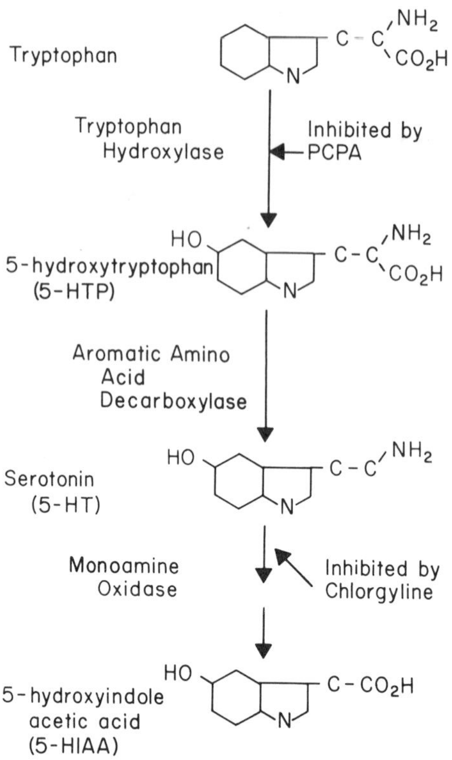

Figure 3. The primary sites of action in the serotonergic biosynthetic pathway of tryptophan, 5-hydroxytryptophan, PCPA, and chlorgyline.

However, 5-HTP can also enter catecholaminergic neurons, be decarboxylated there and serve as a false transmitter (Ng et al., 1977). Like tryptophan, 5-HTP should enhance the frequency of behaviors to which serotonergic systems contribute, yet it is more likely to alter catecholaminergic systems than is tryptophan. When given in combination with PCPA, 5-HTP should partially overcome the decrement in serotonin biosynthesis induced by PCPA alone since the conversion of 5-HTP to serotonin does not require tryptophan hydroxylase. Chlorgyline is a non-competitive inhibitor of monoamine oxidase; it primarily affects the A isoenzyme which preferentially deaminates serotonin and norepinephrine (Johnson, 1968). Thus chlorgyline should increase the neuronal concentrations of serotonin (and norepinephrine) to be released into the synaptic area resulting in augmented stimulation of the postsynaptic receptors. In brief, tryptophan, 5-HTP alone, and chlorgyline treatment should increase serotonergic activity, PCPA should decrease it, and 5-HTP should partially reverse the effects of PCPA.

The results of these drug treatments have been reported in detail elsewhere (Raleigh et al., 1980) and are summarized in Figure 4. The behaviors we monitored included approaching, grooming, resting, eating, locomoting, avoiding, being vigilant, being solitary, huddling, initiating and receiving aggression, and engaging in sexual behavior. Each row of Figure 4 lists the effects of tryptophan, PCPA, 5-HTP, chlorgyline, or PCPA and 5-HTP treatment. For example, grooming was increased by tryptophan, decreased by PCPA, increased by 5-HTP and by chlorgyline; 5-HTP overcame the effects of PCPA treatment. Similarly, approaching was increased by tryptophan, decreased by PCPA, unaffected by 5-HTP, and increased by chlorgyline; 5-HTP overcame the effects of PCPA treatment. A darkened arrow indicates that a behavior changed in the opposite direction to which the arrow points. For example, being solitary was decreased by tryptophan, increased by PCPA, unaffected by 5-HTP, and decreased by chlorgyline; 5-HTP overcame the effects of PCPA.

Before inferring that serotonergic mechanisms contributed significantly to the mediation of a behavior, we required as a minimum criterion that tryptophan altered behavior in one direction and PCPA in the other. Based on that criterion, 8 of the observed behaviors were influenced by serotonergic systems. Approaching, grooming, resting, and eating were augmented by enhancing serotonergic mechanisms, while locomoting, avoiding, being solitary, and being vigilant were diminished. Huddling, initiating and receiving aggression, and engaging in sexual behavior were not consistently affected by these treatments.

We supplemented this multiple drug strategy by determining whether there was a relationship between the frequency with which animals engaged in behavior and peripheral biochemical parameters reported to reflect central serotonergic activity

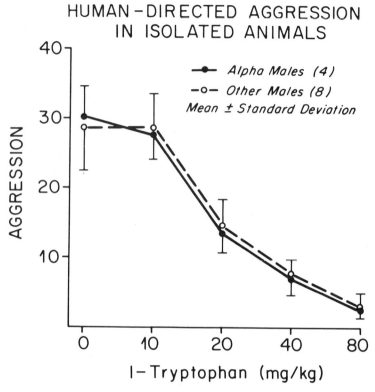

Figure 4. The behavioral effects of the drug treatments. An open triangle indicates an increase in the behavior; an inverted triangle indicates a decrease in a behavior; a circle indicates no change. Closed triangles indicate a behavioral change in the opposite direction. They were used to clarify the pattern of the behavioral changes. All behavioral changes were statistically significant.

in drug-naive animals (Raleigh et al., 1981). The rationale was that if behaviors are indeed influenced by serotonergic mechanisms, and if the peripheral biochemical parameters are related to central serotonergic activity, then in drug-naive monkeys there should be an association between serotonergically-influenced behaviors and the peripheral biochemical parameters. For example, the frequency of approach should be greater in animals exhibiting higher basal levels of these biochemical parameters. In these studies we examined three potential peripheral markers of central

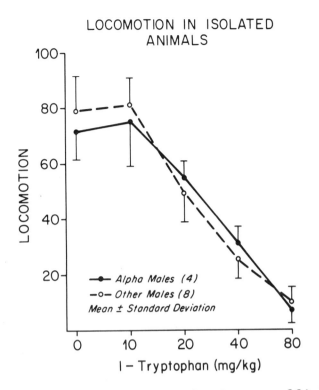

Figure 5. The multiple correlation coefficients of whole blood serotonin, free tryptophan, and total tryptophan with each of the twelve behaviors in all animals together (ALL, n=48), in males only (MALES, n=25), and in females only (n=23). Behaviors monitored included approach (APP), groom (GRM), rest (RST), eat (EAT), locomote (LOC), avoid (AVD), be solitary (SOL), be vigilant (VIG), huddle (HUD), engage in sexual behavior (SEX), initiate aggress (IAG), and receive aggress (RAG). *=p<.05, =p<.01, =p<.001.

serotonergic activity: whole blood serotonin, plasma free tryptophan, and plasma total tryptophan. The rationale for the selection of these has been reviewed elsewhere (Raleigh et al., 1981).

Figure 5 shows the multiple correlations between the three biochemical parameters and the observed behaviors. Considering all animals together, there were significant multiple correlations between the biochemical parameters and all eight of the behaviors inferred to be serotonergically-influenced. For males, there were significant correlations between the biochemical parameters and approaching, grooming, resting, eating, locomoting, being solitary, and avoiding. For females, the multiple correlations were significant for approaching, grooming, resting, avoiding, and being solitary. Considering all animals together, males alone, or females alone, there were no significant multiple correlations for huddle, initiate aggress, receive aggress, or engage in sexual behavior.

Figure 6 shows the partial correlations between the behaviors and each of the biochemical parameters. Whole blood serotonin was inversely correlated with avoiding and being solitary in all animals together and in females alone. In males alone, whole blood serotonin was positively correlated with approaching and initiating aggression and negatively with avoiding. Free tryptophan was positively correlated with approaching, grooming, and eating, and negatively correlated with avoiding and being solitary for all animals together. In males only, approaching, grooming, and eating were positively related to free tryptophan concentrations and locomoting, avoiding, and being solitary were negatively correlated with free tryptophan. In females only, the partial correlations of free tryptophan were positive for approaching and grooming and negative for avoiding. The correlations between total tryptophan and the twelve behaviors were virtually identical to those of free tryptophan. As this figure shows, 35 of the 36 significant correlations involved behaviors which we inferred on the basis of pharmacological data to be serotonergically-influenced. Furthermore, the direction of these partial correlations coincided with what would be expected on the basis of the pharmacological data. The partial correlations were positive for those behaviors which were increased by pharmacologically enhancing serotonergic activity and were negative for those behaviors which were decreased following serotonergic enhancement. Thus the multiple and partial correlational data are compatible with the inferences made on the pharmacological studies.

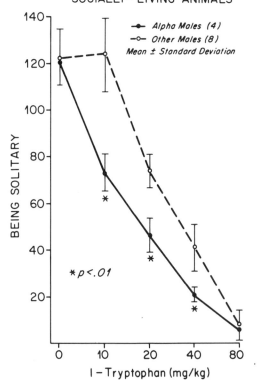

Figure 6. The partial correlation coefficients between whole blood serotonin (BLOOD 5-HT), free tryptophan (FREE TRYP), and total tryptophan (TOTAL TRYP) and the behavioral variables. Symbols are as in Figure 5.

In summary, in our previous work, we have utilized two different approaches, both of which suggest that serotonergic mechanisms contributed to the mediation of a variety of social behaviors. Serotonergic mechanisms appear to enhance approaching, grooming, resting, and eating, and to reduce the frequency of avoiding, locomoting, being solitary, and being vigilant.

STATUS DIFFERENCES IN BEHAVIORAL RESPONSES TO DRUGS

In the course of the experiments described above, we observed that dominant males appeared more responsive to tryptophan than were other males. In dominant males, increases in grooming and approaching appeared earlier and were of a larger magnitude (McGuire et al., in press b; Raleigh and McGuire, 1980). These data provided the initial basis for investigating potential status-related differences in the behavioral responses to drugs directed largely at serotonergic systems.

In these studies we used four established groups of drug-naive vervet monkeys. Each group contained three adult males and three adult females and was maintained in a 3m X 3m X 2m outdoor enclosure. Prior to these studies, all groups had been allowed to stabilize behaviorally for at least three months, during which time animals were observed and habituated to the capture and injection procedures. By the end of this time a dominant male had clearly emerged in each group.

We will report on four behavioral measures. Two were obtained when the animals were in their social groups and two when they were temporarily isolated. The two social behavioral measures were approach and be solitary. These were selected because they are oppositely influenced by drugs affecting serotonergic systems; they do not correlate with each other in baseline conditions; and they occur frequently enough so that drug treatments could be limited to five days. One of the nonsocial measures was the aggressive response to a human stare, a measure that has been used to assess the effects of a variety of biological interventions (Butter and Snyder, 1972). The other nonsocial measure was locomotion, a partial analog to open field tests used in the study of rodents. The observational procedures were identical to those used previously (Raleigh et al., 1980).

Animals were treated with two drugs. One was tryptophan, which as mentioned earlier is a precursor of serotonin. The other was fluoxetine, which is a fairly specific inhibitor of serotonin reuptake (Fuller, 1980) and thereby increases the length of time serotonin is in the synaptic patch. This action should augment the number of post synaptic receptor sites occupied by serotonin, thereby enhancing serotonergic activity. Both tryptophan and fluoxetine were given at four doses (10, 20, 40, and 80 mg/kg daily for tryptophan and

APPROACH IN SOCIALLY-LIVING ANIMALS

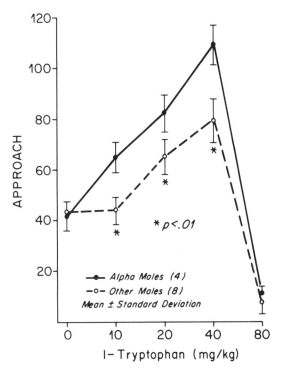

Figure 7. The effects of l-tryptophan on approach. The mean ± (standard deviation) of the number of occurences for four dominant and eight other males is shown. Animals received 0 (saline), 10, 20, 40, and 80 mg/kg of tryptophan per day for a five day period. Differences between the dominant and other males were significant for the 10, 20, and 40 mg/kg treatments.

0.5, 1.0, 2.0, and 4.0 mg/kg daily for fluoxetine). One male was treated at a given time for five consecutive days and other group members received saline. For any animal, at least three weeks intervened between drug treatments. Animals were treated in a randomized order. Behavioral observations began 90 minutes after drug administration and lasted for 90 minutes.

Because fluoxetine and tryptophan produced virtually identical effects on the behavioral parameters, Figures 7 through 10 show only the effects of tryptophan. Figure 7 shows the rate of approaching as a function of 0, 10, 20, 40, or 80

PARTIAL CORRELATION COEFFICIENTS[a]

BLOOD 5-HT		APP	GRM	RST	EAT	LOC	AVD	SOL	VIG	HUD	SEX	IAG	RAG
ALL	(48)	.04	.12	.17	.10	-.12	-.45†	-.40†	.10	.06	.14	.25	-.09
MALES	(25)	.44*	.03	.14	.19	-.06	-.43*	-.31	.02	.03	.29	.43*	-.15
FEMALES	(23)	-.28	.17	-.03	.07	-.16	-.47*	-.52†	.21	.10	.08	.01	-.07
FREE TRYP		APP	GRM	RST	EAT	LOC	AVD	SOL	VIG	HUD	SEX	IAG	RAG
ALL	(48)	.55†	.59†	.10	.43†	-.22	-.56†	-.38*	.09	-.08	-.09	.04	.07
MALES	(25)	.63†	.60†	.29	.52†	-.45*	-.58†	-.43*	.19	-.10	-.13	-.04	.04
FEMALES	(23)	.46*	.53†	.12	.25	-.08	-.49†	-.21	-.02	-.07	-.08	.16	.11
TOTAL TRYP		APP	GRM	RST	EAT	LOC	AVD	SOL	VIG	HUD	SEX	IAG	RAG
ALL	(48)	.56†	.49†	.26	.38*	-.37*	-.45†	-.46†	.08	-.20	.01	.15	.14
MALES	(25)	.61†	.50*	.55†	.39	-.41*	-.48†	-.52†	.17	-.22	-.04	.16	.11
FEMALES	(23)	.46*	.48*	.06	.35	-.34	-.42*	-.29	.01	-.20	.08	.13	.18

Figure 8. The effects of l-tryptophan on be solitary. The mean ± (standard deviation) of the number of occurences for four dominant and eight other males is shown. Animals received 0 (saline), 10, 20, 40, and 80 mg/kg of tryptophan per day for a five day period. Differences between the dominant and other males were significant for the 10, 20, and 40 mg/kg treatments.

mg/kg of tryptophan. Dominant males and other males did not differ in the rate of approach behavior when they were receiving saline. However, at 10 mg/kg, dominant males increased their approach behavior while other males did not. At 20 and 40 mg/kg both dominant males and other males increased approaching but this increase was significantly greater in dominant males. At 80 mg/kg, the soporific effects of tryptophan virtually eliminated approach behavior: both dominant and other males largely sat and rested or slept.

MULTIPLE CORRELATION COEFFICIENTS[a,b]

BEHAVIOR	ALL	MALES	FEMALES
APP	.79‡	.82‡	.67†
GRM	.76‡	.69†	.72†
EAT	.43*	.61†	.61†
RST	.45*	.67†	.49
LOC	.49†	.63†	.41
AVD	.73‡	.81‡	.71†
SOL	.58‡	.62†	.58*
VIG	.47*	.39	.48
HUD	.26	.22	.28
SEX	.21	.34	.18
IAG	.35	.41	.20
RAG	.26	.25	.29

Figure 9. The effects of 1-tryptophan on locomotion in animals temporarily isolated from their social group. The mean $\pm$ (standard deviation) of the number of occurrences for four dominant and eight other males is shown. Animals received 0 (saline), 10, 20, 40, and 80 mg/kg of tryptophan per day for a five day period. There were no differences between the dominant and other males at any dose.

Figure 8 shows that a similar pattern occurred for being solitary. As with approaching, the rate of being solitary was the same for dominant males and other males when they received saline. However, at 10 mg/kg, the dominant males were less solitary while this behavior was unaltered in other males. At 20 and 40 mg/kg both dominant males and other males were less solitary, but the change was larger in dominant males. At 80 mg/kg, the animals were sitting and resting or sleeping but in proximity to other animlas, so being solitary was greatly diminished in both dominant and other males. In brief,

SUMMARY OF THE BEHAVIORAL EFFECTS

BEHAVIOR	TYRPT	PCPA	5-HPT	CHLORG	PCPA-5HTP
	\multicolumn{5}{c}{BEHAVIORAL EFFECTS}				
GROOMING	△	▽	△	△	△
APPROACH	△	▽	○	△	△
REST	△	▽	○	○	○
EAT	△	▽	○	○	○
LOCOMOTION	▲	▼	○	○	○
SOLITARY	▲	▼	○	▲	▼
AVOID	▲	▼	○	▲	▼
VIGILANCE	▲	▼	▼	▼	○
AGGRESSION	○	▼	▼	○	▽
HUDDLE	○	▽	○	○	○
SEX	○	○	○	○	○

Figure 10. The effects of l-tryptophan on human-directed aggressive behavior in animals temporarily isolated from their social group. The mean $\pm$ (standard deviation) of the number of occurences for four dominant and eight other males is shown. Animals received 0 (saline), 10, 20, 40, and 80 mg/kg of tryptophan per day for a five day period. There were no differences between the dominant and other males at any dose.

dominant males differed from other males in that the dose-response curves were shifted to the left for both approaching and for being solitary.

Figure 9 shows the effects of tryptophan on locomotion. Responses were measured when animals were briefly isolated from their social group. For both groups of animals there were dose-dependent reductions in this measure. At 20 and 40 mg/kg the response was reduced, while at 80 mg/kg it was virtually eliminated due to the soporific effects of tryptophan. Unlike the two social measures, there were no status-related differences in the effects of tryptophan or fluoxetine on this measure.

Figure 10 shows that a similar situation occurred for the aggressive behavioral responses to a human stare. While 10 mg/kg had no effect, at 20, 40, and 80 mg/kg there were dose-dependent reductions in this measure. As for locomotion, there were no status-related differences in the effects of tryptophan or fluoxetine on this behavior.

In brief, status-related differences in the effects of tryptophan and fluoxetine were apparent when the animals were tested in their social group. However, status-related differences were not seen when the animals were tested in a nonsocial context.

These data suggest that social status and the environmental test situation are factors which may affect the behavioral responses of vervet monkeys to drugs that enhance serotonergic activity. Because the behavioral distinction between dominant males and other males can be documented quantitatively in captivity and in free-ranging settings, it is unlikely that it is secondary to the stress of living in captivity. Rather this differentiation represents an important dimension of vervet monkey social organization.

When tested in social groups, vervets manifest status-related differences in their behavioral responses to fluoxetine and tryptophan. Since fluoxetine and tryptophan both enhance central serotonergic activity, these behavioral differences are likely to reflect differences between the dominant and other males in the sensitivity of their central serotonergic mechanisms. However, because these drugs may also affect peripheral systems, the possibility exists that differential effects on skin sensitivity, digestion, or intestinal motility may account for these behavioral differences, although such peripheral effects would more likely reduce, rather than enhance approach and other affiliative behaviors. Another, not mutually exclusive possibility is that dominant males and other males are equally sensitive to drug-induced enhancements of central serotonergic systems but that the presence of a dominant male inhibits a nondominant male from expressing the behavioral alterations. Regardless of the behavioral or biochemical mechanisms that account for the effect, social status exerts an unequivocal impact on the behavioral responses to drugs altering serotonergic activity in vervets.

Our observation of no status-related differences when vervets were tested outside of their social group emphasizes the potential utility of using social test situations to monitor the effects of drugs. Slob et al. (1980) and Steklis et al. (1980) have also noted that environmental setting may be important in assessing the behavioral effects of many drugs and hormones. Steklis et al. (1980) demonstrated that in a semi-free-ranging group, progesterone administration to selected females greatly reduces the sexual behavior between males and the treated females. In laboratory settings, where the males' exposure to females is greatly limited, progesterone treatment exerted much less of an effect on sexual behavior. Thus the effects of progesterone treatment observed in a laboratory test situation differ from those seen in a more natural, free-ranging setting. Similarly, the status-related differences we observed in a social group setting were not predictable on the basis of tests conducted in nonsocial settings. These observations as well as other presentations in this volume indicate that documenting the effects of drugs on monkeys in social groups provides information unavailable either from studies of species less closely related to humans or from the examination of monkeys in non-social conditions.

ACKNOWLEDGEMENTS

We would like to acknowledge the technical assistance of Selma Plotkin, Ray Wallace, Mark Vierra, Brian Vierra, and Diane Bushberg. This research was supported in part by the Research Service of the Veterans Administration Medical Center, Brentwood and Sepulveda, in part by grant H80606 from the H. F. Guggenheim Foundation, and in part by grant MH 1646 to the Department of Psychiatry/Biobehavioral Sciences, UCLA. Computing assistance was provided by the UCLA Health Science Computer Center, supported in part by NIH special resources grant Rr-3. Useful suggestions were provided by Lynn Fairbanks, Jeff Flannery, and Michelle Freier.

REFERENCES

Butter, C.N. and Snyder, D.R. Alterations in aversive and aggressive behavior following orbital frontal lesions in rhesus monkeys. Acta Neurobiol. Exp. (Warsz) 32, 525-565 (1972).

Claus, G. and Kling, A. Effects of methaqualone on social-sexual behavior in Macaca mulatta. Paper presented at a Satellite Symposium of the 8th Congress of the International Primatological Society, Pisa, Italy (1980).

Fairbanks, L.A. and Bird, J. Unpublished manuscript.

Fairbanks, L.A. and McGuire, M.T. Inhibition of control role behaviors in captive vervet monkeys (Cercopithecus aethiops sabaeus). Behav. Processes 4, 145-153 (1979).

Fairbanks, L.A., McGuire, M.T., and Page, N. Social roles in captive vervet monkeys. Behav. Processes 3, 335-352 (1978).

Fuller, R.W. Pharmacology of central serotonin neurons. Ann. Rev. Pharmacol. Toxicol. 20, 111-117 (1980).

Jequier, E., Lovenberg, W., and Sjoerdsma, A. Tryptophan hydroxylase inhibitors: the mechanism by which p-chlorophenylalanine depletes rat brain serotonin. Mol. Pharmacol. 3, 274-279 (1967).

Johnson, J.P. Some observations upon a new inhibitor of monoamine oxidase in brain tissue. Biochem. Pharmacol. 17, 1285-1297 (1968).

Keverne, E.B. Social behaviour and its influences on endocrine status. Paper presented at a Satellite Symposium of the 8th Congress of the International Primatological Society, Pisa, Italy (1980).

Koe, B.K. and Weismann, A. p-Chlorophenylalanine: a specific depletor of brain serotonin. J. Pharmacol. Exp. Ther. 154, 499-516 (1966).

Lancaster, J.B. Play-mothering: the relations between juvenile females and young infants among free-ranging vervets (Cercopithecus aethiops). Folia Primatologica 15, 161-182 (1971).

McGuire, M.T. The St. Kitts vervet. Contrib. Primatol. 1, 1-199 (1974).

McGuire, M.T., Cole, S.R., and Crookshank, C.A. Effects of social and spatial density changes in Cercopithecus aethiops sabaeus. Primates 19, 615-631 (1978).

McGuire, M.T., Raleigh, M.J., Yuwiler, A., Brammer, G.L., and Johnson, C.K. Biosociopharmacology I: Basic paradigm and implications, in Sociopharmacology: Drugs in Social Context, C. Chien, ed. Reidel Publ. Co., Dordrecht, Holland (in press a).

McGuire, M.T., Raleigh, M.J., Yuwiler, A., Brammer, G.L., and Johnson, C.K. Biosociopharmacology II: Research options and limitations, in Sociopharmacology: Drugs in Social Context, C. Chien, ed. Reidel Publ. Co., Dordrecht, Holland (in press b).

Moir, A.T.B. and Eccelston, D. The effects of precursor loading in the cerebral metabolism of 5-hydroxyindoles. J. Neurochem. 15, 1093-1108 (1968).

Ng, R.Y., Chase, T.N., Colburn, R.W., and Kopin, I.J. Release of 3-H-dopamine by L-5-hydroxytryptophan. Brain Res. 45, 499-505 (1977).

Oldendorf, W.J. Brain uptake of radiolabeled amino acids, amines, and hexoses after arterial injection. Am. J. Physiol. 221, 1629-1639 (1971).

Raleigh, M.J., Brammer, G.L., Yuwiler, A., Flannery, J.W., McGuire, M.T., and Geller, E. Serotonergic influences on the social behavior of vervet monkeys (Cercopithecus aethiops sabaeus). Exp. Neurol. 68, 322-334 (1980).

Raleigh, M.J., Flannery, J.W., and Ervin, F.R. Sex differences in behavior among juvenile vervet monkeys (Cercopithecus aethiops sabaeus). Behav. Neural. Biol. 26, 455-465 (1979).

Raleigh, M.J. and McGuire, M.T. Biosocial pharmacology. MacLean Hospital J. 5, 73-86 (1980).

Raleigh, M.J., Yuwiler, A., Brammer, G.L., McGuire, M.T., Geller, E., and Flannery, J.W. Peripheral correlates of serotonergically-influenced behaviors in vervet monkeys (Cercopithecus aethiops sabaeus). Psychopharmacology 72, 241-246 (1981).

Redmond, D.E. Social effects of alterations in brain noradrenergic function on untreated group members. Paper presented at a Satellite Symposium of the 8th Congress of the International Primatological Society, Pisa, Italy (1980).

Rowell, T.E. The social organization of caged groups of Cercopithecus monkeys. Anim. Behav. 19, 625-645 (1971).

Sassenrath, E.N. Studies in social adaptability in rhesus monkeys: Environmental, experiential, and pharmacological influences. Paper presented at a Satellite Symposium of the 8th Congress of the International Primatological Society, Pisa, Italy (1980).

Slob, A.K., Schenk, P.E., and Nieuwenhuijsen, H. The effects of cyproterone acetate on social and sexual behavior in the adult male laboratory housed stumptail macaque (Macaca arctoides). Paper presented at a Satellite Symposium of the 8th Congress of the International Primatological Society, Pisa, Italy (1980).

Steklis, H.D., Linn, G., Howard, S., Harvey, N., Kling, A., and Tiger, L. Hormonal and social-environmental influences on sexual behavior of stumptail macaques (Macaca arctoides). Paper presented at a Satellite Symposium of the 8th Congress of the International Primatological Society, Pisa, Italy (1980).

Struhsaker, T.T. Social structure among vervet monkeys (Cercopithecus aethiops). Behaviour 29, 83-121 (1967).

Yuwiler, A. Conversion of d- and l-tryptophan to brain serotonin and 5-hydroxyindoleacetic acid and to blood serotonin. J. Neurochem. 20, 1099-1109 (1973).

Chapter 5
Progesterone and Socio-Sexual Behavior in Stumptailed Macaques *(Macaca Arctoides):* Hormonal and Socio-Environmental Interactions

H.D. Steklis, G.S. Linn, S.M. Howard, A. Kling, and L. Tiger

INTRODUCTION

Evidence from several Old World monkey species indicates that heterosexual interactions are affected by endogenous cyclical variations in or experimental manipulations of the female's hormonal status. Laboratory male-female pair test studies have shown that in Macaca fuscata (Enomoto et al, 1979) and M. mulatta (Michael and Bonsall, 1979), for example, male frequency to ejaculation is significantly higher around the middle part of the female menstrual cycle, when ovulation is most likely to occur, than during other parts of the cycle. Variation in female behaviors, such as midcycle peaks in frequency of sexual invitation and approach to the male (e.g., in M. fuscata, Enomoto et al, 1979) or in spending more time in proximity to the male (e.g., in M. mulatta, Czaja and Bielert, 1975), has also been reported for both laboratory pairs (Keverne, 1976, for review) and females living in large heterosexual groups (e.g., Cochran, 1979). Furthermore, a variety of studies have shown that high levels of female sexual attractivity, receptivity, and initiative in soliciting copulation (or proceptivity, cf. Beach, 1976) are related to high plasma concentrations of estradiol (probably in combination with testosterone), while decreased male initiated sexual behavior (i.e., decreased female sexual attractiveness) is related to high plasma concentrations of progesterone (Michael and Bonsall, 1979, for review).

While these studies suggest that sex steroids have a significant influence on heterosexual interactions, a growing body of evidence indicates that a variety of non-hormonal factors interact with and often modify hormonal influences on sexual behavior. For example, it has been apparent in both earlier (e.g., Herbert, 1967) and more recent (e.g., Zumpe and Michael, 1977) investigations of rhesus monkey sexual behavior that high variability may exist between subjects in the extent to which female hormonal status influences heterosexual behavior. Differences between individuals in preferences for particular sex partners (see Keverne, 1976, for review) or in sensitivity to female hormonal status (cf. Herbert, 1967) are among the factors likely to account for this type of variability.

Socio-environmental factors or testing conditions also may greatly modify the degree to which hormones mediate or correlate with behavioral events. In pigtail macaques (M. nemestrina), for example, there was no relationship between certain female proceptive behaviors (i.e., presenting or proximity responses) and phase of the menstrual cycle or hormone levels during pair tests (Eaton and Resko, 1974); however, these behaviors did vary with phase of the menstrual cycle when three females were tested with a male (Goldfoot, 1971). Similarly, in stumptailed macaques (M. arctoides), no significant variation in either male or female initiated sexual activity over the course of the female's menstrual cycle has been reported when a male is tested with one female (Slob et al., 1978a) or with two females (Slob et al., 1978b). In a large, stable heterosexual group, however, significant midcycle rises in female attractiveness and proceptivity were observed (Harvey, this volume). It seems, therefore, that the complexity of the social setting is an important factor in the expression of sexual behavior in several non-human primate species (see also Johnson and Phoenix, 1978; Cochran, 1979).

Collectively, these studies suggest that variations in social conditions may greatly modify the influence of hormones on sexual behavior in several macaque species. In this chapter we will review the results of three studies which were conducted to examine how variations in socio-environmental setting may alter the behavioral consequences of long-term treatment of female stump-tailed macaques with medroxyprogesterone acetate (Depo-Provera, MPA). In the first study the effects of MPA on socio-sexual behaviors were examined in a stable, semi-free-ranging heterosexual group. In the second study, the effects of MPA were evaluated in time-limited behavior tests of several one male-two female trios. In the final study, such time-limited behavior tests were also utilized in order to further explore the effects of MPA on the animals from the first study under different test conditions. Thus, one male and two females who had previously served as subjects in the social group constituted the test trio.

STUDY ONE

The aim of this study was to determine the effects of a long-acting progestational agent (MPA, Depo-Provera) on sexual and related social behaviors in a stable social group of stumptailed macaques. Although this compound has been used in human females as an effective alternative means of contraception (in place of orally administrated agents) (e.g., Rosenfield, 1974), its effects on male-female socio-sexual interactions have not been investigated.

There are a variety of studies on nonhuman primates, however, which indicate that progesterone is related to decreased female sexual attractiveness. For example, daily intramuscular administration of progesterone to ovariectomized rhesus females, which produced plasma concentrations equivalent to endogenous luteal peaks, resulted in a significant decline in mounting rate and ejaculation by male partners (Baum et al., 1977). Since similar behavioral effects were obtained by administering progesterone intravaginally without a significant increase in plasma concentration, it was suggested that progesterone may lower female sexual attractiveness by blocking the estrogen-induced production of a vaginal factor (e.g., a pheromone) which normally enhances sexual attractiveness.

The role of progesterone in female initiated sexual activity is less clear in that various studies have produced conflicting results (see Baum, 1978, for review). Baum et al. (1977) reported an increase in the display of sexual invitations by progesterone treated, ovariectomized females toward their male partners, whereas an earlier study (Zumpe and Michael, 1970) failed to report any changes in proceptivity. Some of this inconsistency in behavioral results is probably due to the choice of behavioral measures commonly used to reflect proceptivity. For example, female "presenting," a frequently used measure, also occurs in non-sexual contexts (e.g. dominance, agonism), and hence its frequency of occurrence may be influenced by the non-sexual aspects of the social relationship. In pair tests, therefore, female initiated behaviors may more often reflect the status of male-female social relations than female sexual motivation. Zumpe and Michael (1977) have shown, for example, that in rhesus monkey pairs the frequency of female proceptive behavior varies according to the frequency of male ejaculatory behavior.

Johnson and Phoenix (1978) have suggested that in settings where females have the most control over the completion of copulation, they are most likely to show variations in proceptivity linked to variations in hormonal status. This may explain why in circumstances where female behavior is less directly influenced by male partners, progesterone has been more consistently related to decreased proceptivity. When an operant schedule requirement was employed, requiring a rhesus female to lever press to release a male partner, indicating female sexual motivation, a luteal depression in release latency was evident (Bonsall et al., 1978). Similarly, in a large social group, female rhesus monkeys exhibited an abrupt cessation of sexual invitations during the early luteal period (Cochran, 1979). A more direct relationship between alterations in female sociosexual behaviors and progesterone was demonstrated in a study of free-ranging baboons (Papio ursinus) (Saayman, 1973). Females were given progesterone implants which were sufficient to induce deturgescence of the sex skin. Females with active Provera implants exhibited decreased proximity to males, groomed males less and females more frequently than when compared to periods when release of Provera from the implants declined. Furthermore, these resultant changes in heterosexual interaction patterns were identical to those observed during the luteal phase of normally cycling females.

For humans there are some reports, though less consistent and generally more difficult to interpret (e.g., see Brancroft, 1978, for review), which suggest that progesterone contributes more to decreased female than to male initiated sexual activity. In one survey (Grant and Mears, 1967), women taking contraceptive pills high in progesterone were far more apt to complain of "loss of libido" than those taking preparations high in estrogen. This finding, perhaps explainable on the basis of non-specific depressive mood changes (cf. Grant and Pryse-Davies, 1968), is nonetheless consistent with results of a study (cited in Bancroft, 1978) conducted by the Royal College of General Practitioners using 20,000 pill users and the same number of control subjects. Analysis of questionnaires revealed that complaints of decreased sexual interest were more than four times as common among women taking oral contraceptives than among women employing other means of birth control. Although Udry and Morris (1970) found that use of contraceptive pills was associated with elimination of the luteal depression in sexual activity reported earlier by non-pill users (Udry and Morris, 1968), in these studies no distinction was made between male-initiated and female-initiated sexual activity. In a more recent study where this distinction was made (Adams et al., 1978), it was found that the midcycle peak in the frequency of both autosexual and female-initiated heterosexual activity established for non-pill subjects was abolished in women taking low estrogen contraceptive pills. Male-initiated sexual activity, however, did not vary between groups.

In summary, studies in nonhuman primates have linked progesterone to both decreased female sexual attractiveness and, in situations where females have substantial control over mating activity (e.g., a social group), decreased proceptivity. In humans, progesterone may decrease female-but not male-initiated sexual activity. Our aim in the present study, therefore, was to determine whether MPA would have similar influences on socio-sexual behavior in a stable, social group. The results reported here are preliminary in that only data on copulation and grooming are presented. Details of the methods and results have appeared elsewhere (Steklis et al., 1982).

Methods

The subjects for this study were one vasectomized adult male, a two and one-half year old juvenile male, and eight adult female stumptailed macaques. The animals were part of a social group of 15 animals established on the .70 hectare island facility (C.R. Carpenter Primate Center) in Bermuda (see Esser et al., 1979, for further details). The experiment was conducted over the course of 424 days, divided into four phases: Phase 1 (118 days) consisted of pre-treatment behavioral observations; at the beginning of Phase 2 (132 days) four adult females were treated with a single IM injection of 30 mg MPA (Depo-Provera, Upjohn); at the beginning of Phase 3 (84 days) the second four adult females were given this dosage of MPA, and one female was re-treated with it; at the beginning of Phase 4 (90 days), all females again received MPA with those treated during Phase 2 receiving 100 mg MPA and the four treated during Phase 3 receiving a second 30 mg injection. The 30 mg dose was selected because it had been shown to be effective in suppressing estradiol surges in rhesus monkeys to values either below the range of normal menstrual cycles or similar to those of early follicular phases (K. Kirton, Upjohn, Kalamazoo,Michigan, personal communication). The higher dosage of MPA was used to detect dose-dependent behavioral effects.

Grooming interactions between females and between males and females were recorded during all phases by using a focal subject technique (Altmann, 1974). Copulation, which consisted of mounting, intromission, pelvic thrusting, and ejaculation accompanied by species-typical postural, facial, and vocal signals (Chevalier-Skolnikoff, 1975) was recorded on an all-occurrence basis. Observations were conducted on an average of six days per week and were balanced for time of day. Occasional blood samples were obtained from treated females for serum RIA of MPA as described previously (Steklis et al., 1982).

Results and Discussion

Behavioral data were analyzed by computing mean rates of copulation/observation day/phase received by each female and by computing mean grooming rate/hour of focal subject observations/phase for each subject. Rates of mounting, intromission, or pelvic thrusting were not analyzed independently of ejaculation, as in only four instances mounting was not followed by intromission, pelvic thrusting, and ejaculation. The effect of MPA treatment on these behavioral variables was determined by a single-factor (i.e., treatment with three levels: 0 mg MPA, 30 mg MPA, 100 mg MPA) General Linear Models Analysis of Variance (ANOVA) for repeated measures. In this analysis the 0 and 30 mg treatment groups contained values representing repeated measures on the same subject. In order to follow temporal changes in behavior across phases, a phase factor with subject as a repeated measure was also examined. Data for seven instead of eight female subjects were included in the analysis of copulation rates, as one female, who had given birth prior to the onset of the study and was lactating, was not observed to engage in any copulations during the course of the study.

Figure 1 summarizes the copulatory history of each female throughout the entire course of the experiment. Following the first injection of 30 mg MPA, females six, nine and 10 ceased copulating within the first day and were not observed to copulate again for approximately 10 weeks, whereas

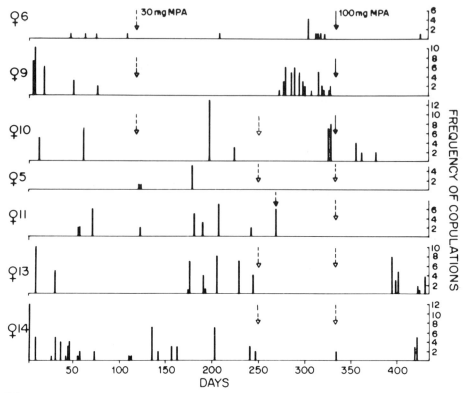

Figure 1. Copulatory history of female subjects.

copulations were recorded during this period for untreated females five, 11, 13, and 14. It is noteworthy that no previous copulations had been recorded for female five. This differential effect on copulation frequency of treated vs. untreated females was replicated following treatment of previously uninjected females five, 11, 13, and 14 with 30 mg MPA. Again, copulations were not observed for any treated female for approximately 60-80 days post-injection, while females six and nine, now untreated, did engage in copulation. This elimination of copulation for a minimum of 60 days post-injection was again observed for six of seven treated females during the final phase of the experiment. Moreover, during this period, when all females had received MPA, three of the four cases of copulation without ejaculation occurred, twice with female 10 and one with female 14.

When rates of copulation for treated and untreated females are compared during phases two and

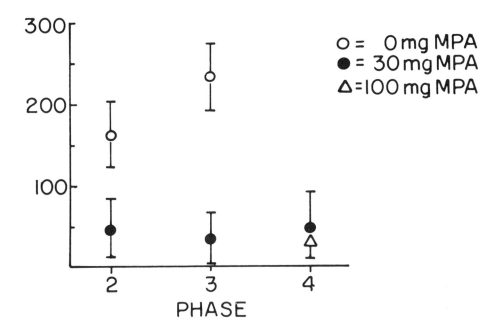

Figure 2. Changes in rates of copulation for each treatment group during treatment phases. All values are expressed as a percentage of phase 1 (pre-treatment) values.

three, it is evident that rates for treated females decreased substantially while rates for untreated females increased relative to phase one rates (see Figure 2). The ANOVA revealed significant differences between mean rates of copulation for females receiving 0 mg, 30 mg, and 100 mg MPA, and a post hoc comparison of the means (Duncan's multiple comparison test) showed a significant (p-0.01) difference between the 0 mg group and either the 30 or 100 mg groups; however, there was no signficant difference between the 30 and 100 mg treatment groups. Furthermore, although the total copulation rate tended to decline across phases (see Table 1), differences between phases were not significant.

In contrast to these changes in copulatory behavior, grooming rates between group members were unaffected by MPA treatment, nor were there any significant changes across phases (Table 1).

TABLE 1

RATES OF COPULATION AND GROOMING AFTER MPA TREATMENT*

	Copulation	Grooming			
		Female		Male	
		Initiated	Received	Initiated	Received
Treatment Group (mg MPA)					
0 (n=15)	0.18+0.04	5.73+0.90	4.60+1.02	0.28+.15	1.15+0.40
30 (n=13)	0.06+0.03	5.61+0.89	4.78+0.83	0.10+.05	1.03+0.30
100 (n=4)	0.03+0.02	5.89+3.53	5.64+3.32	0.28+.13	1.17+0.83
Phase					
1	0.14+0.05	5.39+1.10	4.73+1.56	0.06+0.04	0.77+0.29
2	0.12+0.04	4.64+0.87	3.69+1.04	0.22+0.12	1.11+0.36
3	0.12+0.07	7.21+1.44	5.86+1.23	0.39+0.27	1.74+0.72
4	0.06+0.04	5.56+1.85	4.93+1.70	0.16+0.08	0.79+0.42

*Grooming data are expressed as mean $\pm$ SEM rate per hour of focal subject observations. Copulation data are given as mean $\pm$ SEM rate per observation day.

The results of the serum MPA RIA's indicate that MPA reached its highest concentration within two weeks, and declined to minimally detectable concentrations within 90 days.

In summary, the results of this study show that MPA has a specific influence on patterns of copulatory activity in a stable social group: Copulations between males and treated females were eliminated for approximately 10 weeks after treatment (which roughly corresponds to the time that MPA was detectable in serum), while copulations involving untreated females continued or were initiated (in some cases for the first time) during these periods.

The data are therefore in agreement with those from other species (reviewed above) on the effects of progesterone on sexual attractiveness; they do not, however, concur with findings from previous studies which failed to demonstrate effects of progesterone on any aspect of sexual behavior in laboratory-tested stumptailed macaques (Slob et al., 1978b). We therefore decided to evaluate the possiblity that behavioral testing conditions (i.e., a stable social group vs. time-limited laboratory tests) affect the behavioral outcome of MPA treatment in this species.

STUDY TWO

A major aim of this study was to examine the effects of MPA treatment on socio-sexual behavior of several one male-two female trios that were tested in a manner similar to that used in previous laboratory studies of this species (cf. Slob et al., 1978b). This would permit a closer comparison to the results of previous investigations than was possible in Study One. In addition, moving this investigation into the laboratory allowed us to track on a regular basis female menstrual condition and serum concentrations of progesterone, measures which were not obtainable on a regular basis in the island colony. A detailed presentation of the methods and results of this study have also appeared elsewhere (Steklis et al., 1982).

Methods

Four tubal-ligated adult female and three adult male laboratory-born stumptailed macaques with extensive previous social experience served as subjects. One of the males had been castrated early in adulthood and was maintained on subcutaneous testosterone pellets (75 mg) (Oreton, Schering) replaced at six to eight week intervals throughout the study. Analyses of serum concentrations of testosterone (obtained at regular intervals by RIA) and copulatory behavior showed the castrated male's values to be within the range of variation for the two normal males.

In brief, the design and procedure of the study were as follows: Two pairs of females were formed on a random basis, and each pair was observed with each male during a 30 minute test on three or four occasions per week. After 12.5 weeks of testing, the females in each pair that received the higher mean rate of ejaculation from at least two of the three males was designated the "preferred" female and was injected with 30 mg MPA IM. Testing was then resumed for an additional 17 weeks. At this time a third pair was constituted, consisting of the two previously untreated females, and the same set of adult males was tested with pair three as previously. After five weeks, the preferred female of this pair was treated with 30 mg MPA and testing was resumed for 4.5 weeks. Vaginal swabs were obtained daily, and the onset of bleeding noted. Blood samples were taken two to three times a week for serum progesterone RIA's as described previously (Steklis et al., 1982).

For the test sessions, individually caged subjects were brought in a transfer cage to a 1 x 2 x 1.5 m lexan chamber, where they were observed by one of three observers from behind a one-way mirror. Eleven sexual behavior and four social behavior measures were recorded using an all-occurrence method. The behaviors were selected and defined according to previously published descriptions for this species (e.g., Goldfoot et al., 1975), and included frequencies of genital exploration, mounting, intromission, ejaculation, latency to first ejaculation, post-ejaculatory intromission duration, length of inter-ejaculation interval, frequencies of male-induced and spontaneous female presentation, female reachback during the mount, withdrawal by the female in response to sexual contact, frequency of male to female groom solicitation, duration of female to male grooming, frequencies of female to male fear grimacing, and male to female threat or physical aggression.

Results and Discussion

The effects of MPA treatment on each of the behavioral variables (i.e., rates or durations/test) were determined by a two-factor (i.e., treatment group and pre-/post-treatment) ANOVA for repeated measures, in which a significant (p-.05) phase x treatment group interaction was considered as indicative of a treatment effect. Additional post hoc comparisons were made by use of dependent Student's t tests.

Of the 15 behaviors measured, only ejaculation rate and duration of female to male grooming were altered significantly by MPA treatment (see Table 2). Mean rates of ejaculation received by treated females decreased while rates for untreated females increased in the post-treatment phase compared to pre-treatment phase rates. Furthermore, the pre-/post-treatment comparison of mean ejaculation rates for all male-female test trios showed that the mean decrease for treated subjects was significant ($t(8)=2.26$, $p<0.05$) as was the mean increase for untreated subjects ($t(8)=6.32$, $p<0.001$). A further comparison between overall mean rates of ejaculation received by treated and untreated females from the nine possible male-female trios demonstrated significant differences in both the pre-treatment ($t(8)=2.18$, $p<0.05$) and post-treatment phases ($t(8)=2.18$, $p<0.05$). Means for females in the treated group were lower than those in the untreated groups during the post-treatment phase, whereas pre-treatment mean rates were higher for females assigned to the treatment group than for females assigned to the untreated group. This indicates

TABLE 2

EFFECT OF MPA ON EJACULATION AND GROOMING IN MALE-FEMALE TRIOS[A]

Females	Ejaculation Received		Grooming of Male	
	Pre-treatment	Post-treatment	Pre-treatment	Post-treatment
Treated (30 mgMPA)	1.75±0.40+	1.15±0.19+*	14.50±2.63+	9.82±1.98**
Untreated	0.80±0.28	1.81±0.31**	4.04±2.26	6.93±1.46*

[A]Ejaculation data are given as mean ± SEM rate/test. Grooming data are given as mean ± SEM duration/test. Means given are for 9 possible male-female trio combinations.

+P<0.05 difference between values for treated and untreated females (Student's dependent t test)

*P<0.05 difference between pre- and post-treatment values (Student's dependent t test)

**P<0.01 difference between pre- and post-treatment values (Student's dependent t test) that MPA treatment produced a switch in ejaculation partner preference. Examination of ejaculation rates for individual male-female test trios showed considerable variation in this respect, with a reversal in partner preference occurring in six of nine trios.

As shown in Table 2, the effects of MPA on female to male grooming paralleled those for ejaculation; Treated females groomed males less, while untreated females groomed males more relative to pre-treatment durations. Unlike ejaculation, however, the means for grooming were not significantly different between treated and untreated females in the post-treatment phase.

These behavioral changes were accompanied by alterations in serum concentrations of progesterone and vaginal bleeding. The normal post-ovulatory rises in serum progesterone concentrations observed in all untreated females were abolished for approximately 100 days, and vaginal bleeding was not evident for 55-85 days post-injection.

This study shows that in laboratory trios, MPA reduced ejaculation rates with treated females while increasing rates with untreated females, a change that was accompanied by an inhibition of ovarian activity in treated females. Unlike in Study One, however, ejaculation was not at any time abolished with treated subjects nor were copulatory behaviors other than ejaculation (e.g., mounting) affected by MPA treatment. Moreover, female to male grooming was altered in the test trios but not in the social group. These differences in the behavioral consequences of MPA treatment between the two studies suggest that test conditions modify the behavioral effects of MPA. Specifically, they point to the amount of socio-sexual contact between males and females as one potentially important variable. Thus, limited socio-sexual contact could affect male sensitivity to changes in female ovarian status. For example, Slob et al., (1978b) tested males with progesterone treated females during 15 minute sessions once per week, which is substantially less than the 60-90 min/week of contact provided for males with a particular female pair in this study. This limited contact may account for the lack of effect on heterosexual interactions reported by Slob et al. In a final study we therefore examined the possibility that reduced heterosexual contact alters male copulatory responses to MPA treated females.

STUDY THREE

In order to compare in the same animal the effects of varying test conditions (i.e., social group vs. time-limited trio tests) on the behavioral outcome of MPA treatment, we decided to re-test the vasectomized adult male and two adult females (numbers 14 and 05) that had previously served as subjects in the semi-free-ranging group. Re-testing the same subjects is of some importance as it reduces the likelihood of between subject (i.e., especially male) variation accounting for differences in results between the previous two studies and between our data and those of other investigators. Our prediction was that a male given access to two familiar copulatory partners only once per week for 30 minutes would not display a cessation in copulation with the treated member of the female pair.

Methods

The vasectomized adult male from the semi-free-ranging group described in Study One was placed in an outdoor 2.5 x 2 x 1.5 m wire mesh cage contained within a 3.5 x 3.5 x 3 m wire mesh cage. Two sides and the top of the inner cage were partially covered to provide shelter. A mesh covered wood partition could be operated from outside the enclosure. This double cage design allowed visual and vocal contact between the male and the rest of the group, yet restricted other forms of contact to test sessions.

Purina monkey chow, supplemented with fresh fruit and naturally occurring forage were fed daily, and water was available ad libitum. The two adult females remained with the group at all times except for test sessions.

Testing began with a procedure similar to that employed in Study Two, except for test frequency, which was designed to be weekly, approximating the male test frequency utilized in earlier studies (i.e., Slob et al, 1978b). Prior to each session the two females were caught in chow-baited trapping cages and subsequently lured into a transfer cage with fruit rewards. Due to this method of capture, which avoided rough handling of individuals, tests did not always occur precisely at weekly intervals. Hence, although there was always a minimum of one week between test sessions, the interval between any two successive tests ranged from 7-14 days.

Following capture of the females, the partition of the inner cage was closed, restricting the male to one side, and the two females were intoduced to the other side. The male was then allowed to enter for a thirty minute test period. Behavioral scoring techniques for both periods were identical to those already described for Study Two. The majority of the observations were made by one observer who also participated in Studies One and Two. An additional observer was utilized, and the range of interobserver reliability for all behavior categories was 89-97%.

Following completion of each test session the male was again restricted to one-half of the inner cage; females were released and rejoined the group.

Following eight test sessions the sexually preferred female was determined according to number of copulations received and treated with 30 mg MPA IM. This was followed by a series of seven post-treatment test sessions.

Results and Discussion

The results of the test sessions are summarized in Table 3. Data were analyzed using Student's t tests for independent samples.

During the pre-treatment phase, all mounts and ejaculations occurred with one female (#14), who also received significantly more genital explores ($t=11.90$, $p<.01$) and initiated more male induced presents ($t=10.52$, $p<.01$). As 14 was clearly the

TABLE 3

EFFECT OF MPA ON SEXUAL BEHAVIORS IN MALE-FEMALE TRIO ISOLATED FROM SOCIAL GROUP[A]

Female	Ejaculation		Mounting		Genital Exploration		Male-Induced Present	
	Pre-treatment	Post-treat.	Pre-treatment	Post-treat.	Pre-treatment	Post-treat.	Pre-treatment	Post-treat.
14	5.25±0.25+	3.0±0.83(T)*	5.25±0.25+	3.0±0.83(T)*	6.0±0.27+	3.7±0.49(T)**	6.0±0.27+	3.0±0.61(T)**
05	0	2.70±0.96	0	3.28±1.18*	1.0±0.32	3.42±1.15	1.60±0.34	4.0±1.0*

[A]Data are given as mean ± SEM rate/test. Means are for 8 pre- and 7 post-treatment tests.

(T) = value following treatment with 30 mg MPA

+$p < 0.05$ difference between subjects (Student's unpaired t test)

*$p < 0.05$ difference between pre- and post-treatment values (Student's unpaired t test)

**$p < 0.01$ difference between pre- and post-treatment values (Student's unpaired t test) preferred copulatory partner, she was treated with 30 mg MPA IM.

Treatment with MPA failed to produce significant differences between females in any of the behavioral variables measured. Although there was a decline in ejaculations received by the treated female (t=2.58, p<.05), and an increase in ejaculations received by the untreated female (t=2.8, p<.05), there was no difference between females in ejaculations received. Similarly, there was no difference between females in genital explores received, male induced presents, or mounts received, although in some cases there were significant changes for individuals between phases.

During each post-treatment test session, the first copulation always occurred with female 14. If copulations took place with 05 during a test session, the remainder of the copulations during that session would be with her; however, copulations did not occur with 05 during each test session.

This study further supports the position that socio-environmental conditions (i.e., social grouping, social and sexual contact) are important variables when assessing the influence of progestins on sexual behavior in M. arctoides.

In contrast to the semi-free-ranging situation, where MPA treatment of females brought about a cessation of copulation for more than 60 days with treated but not untreated females, copulations were still regularly observed with the MPA treated females, and no preference switch occurred when contact was limited to weekly tests. As in Study Two, MPA treatment of the preferred female was associated with a decline in the number of copulations she participated in, while copulations increased for the untreated female; however, this difference between the females did not reach sigificance. These results, therefore, suggest that the differences observed in the behavioral results of MPA treatment in Study One and Study Two are the consequence of reduced socio-sexual contact between males and females, rather than the result of variation among males in their response to MPA treated females.

GENERAL DISCUSSION

These studies indicate that differences in socio-environmental setting account for the observed variation in the behavioral consequences of MPA treatment. In the semi-free-ranging heterosexual group MPA was effective in eliminating copulation for at least 60 days with treated females during periods when untreated females were present. Isolation of one male for testing with two females from the group modified this effect in that while the copulation rate with the treated female was reduced post-treatment, during each post-treatment test at least one copulation with the treated female was observed. The modifying influence of test conditions was further confirmed by the results of Study Two.

In keeping with the results from other laboratory studies on the behavioral effects of progesterone administration (see Introduction), the primary effect of MPA appears to be one of lowered female sexual attractiveness. Although in all trios tested the overall mean ejaculation rates were significantly reduced for treated females, there was variation between subjects in the effects of MPA on both the range of and degree to which measures of sexual attractiveness were altered. Thus, in the trio composed of members of the Study One social group, significant reductions for treated females were also noted in mounting and genital exploration rates received. Similarly in three of the nine possible male-female trio combinations of Study Two, significant mean decreases occurred in rates of mounting and genital exploration initiated with treated females.

An examination of the pre- to post-treatment changes in ejaculation rates for individual males with each of the three female pairs, showed that although there was no significant variation between males in the effect of treatment on rates with treated vs. untreated pair members, post-treatment rates for treated females did not decrease in all trio combinations. (It should be noted that post-treatment ejaculation rates for untreated females, by contrast, did increase in all nine trio combinations.) In only four of nine trio combinations was the post-treatment decrease relative to the pre-treatment mean sufficient to avoid overlap of the standard errors.

These observations suggest that variation in the social dynamics of particular trio combinations interacts with the effects of MPA treatment on sexual attractiveness. It is clear from the present set of studies, for example, that males sexually preferred one member of each female pair prior to treatment. In Study Two, all three males showed higher mean ejaculation rates/test with the same female in eight of nine possible trio combinations (although in only 5 of 8 was this preference significant, $p<0.05$, as determined by a separate post hoc chi-square analysis of each trio combination). A sexual preference was also evident in the Study Three trio and in the social group where some females received higher frequencies of copulations than did others (see Figure 1). Thus there are differences among females in their attractiveness to males not only within a test trio but also between test trios, such that MPA treatment is likely to reduce the sexual attractiveness of some females more than that of others. Preferred sexual associations of this sort have been documented in many previous studies of both laboratory-housed trios (e.g., M. mulatta, reviewed by Keverne, 1976) and free-ranging groups (e.g., M. fuscata, Wolfe, 1979); however, the contributions of physiological-behavioral factors to this type of variation between females in sexual attractiveness are at present unclear.

Measures of female receptivity or proceptivity were not consistently altered by MPA treatment in the test trios. Male-induced presenting, a measure commonly used as an index of female receptivity, was significantly lowered in the treated female of the Study Three trio, and lower also (i.e., no overlap between standard errors of the means) in the treated female of one of the three female pairs used in Study Two in tests with two of the three males. This, again, suggests a strong interactive effect between social dynamic factors (e.g., social or sexual preference, dominance relations) and hormone treatment, so that it becomes difficult to determine which is responsible for the behavioral changes observed.

In the trios used in Study Two, female to male mean groom durations were also affected by MPA treatment, as treated females showed a post-treatment reduction while untreated females showed an increase. In the Study Three trio grooming did not occur with sufficient frequency to permit statistical analysis; however, in the social group no aspect of grooming behavior was altered by treatment. It is likely that this difference between the social group and the test trios is not a consequence of a direct differential effect of MPA on grooming per se in two distinct settings, but rather reflects a particularly close association between grooming and copulation in the trios. Thus, grooming generally occurred between sex partners and was limited to periods immediately preceding and following ejaculation. Attempts by the female who was not a sex partner to groom the male were usually discouraged by joint threats and/or aggression from the male and his sex partner. Therefore, these alterations in grooming in the trios are likely to be a secondary consequence of the primary changes in female attractiveness.

In the social group, opportunities for females to groom males were probably not as restricted (or as exclusively tied to sexual interactions) as in the time-limited trio tests; therefore, grooming in the group setting is more of an additive index of non-sexual and sexual relationships. In a study of consort and grooming relationships in the free-ranging Arashiyama West troop of the closely related Japanese macaque (M. fuscata) it was found that although year long grooming involved pairs of monkeys that were significantly different from those for consorting, patterns of grooming frequencies were not disrupted during the mating season (Baxter and Fedigan, 1979). Thus, in the stable social group of stumptailed macaques studied here, grooming may reflect a diversity of mutually exclusive social bonds that are maintained despite changes in sexual associations due to MPA treatment. This view further suggests that in free-ranging baboons there is a closer connection between grooming and consort bonds than there is in some macaque species, since luteal phase as well as progesterone implanted females showed decreased frequencies of male grooming in association with decreased male sexual interest in them (Saayman, 1973).

The specific effect of a reduction in sexual attractiveness of MPA treated females appears to be mediated through the actions of this hormone on ovarian and reproductive tract status. Hormonal evidence from the present studies indicates that MPA suppresses normal ovarian function (i.e., as evidenced by the low, non-cyclical serum progesterone concentrations) for about 100 days, which is in the lower part of the range of suppression reported for rhesus monkeys after a 150 mg injection (Mora and Johanssen, 1976) and comparable to the inhibition of ovulation noted in baboons following 50 mg IM (Fotherby and Goldzieher, 1980). The time course for serum MPA concentrations for subjects of Study One to reach undetectable levels was also comparable (i.e., 90 days) to that reported in the latter study on baboons. MPA thus reduces estrogenic actions on the reproductive tract, which may alter its olfactory (and perhaps tactile) qualities normally associated with sexual attractiveness (e.g., in M. mulatta, Michael and Keverne, 1968; Baum et al., 1977). In addition to this effect, which is secondary to the inhibition of follicle development, MPA apparently also has a direct action on reproductive tract tissue, in that reduced amounts of cytoplasmic estrogen receptor were found in the uterine cervix of women treated with this agent (Rall et al., 1978). Similar alterations were also reported following progesterone administration in rhesus and cynomolgous macaques (M. fascicularis) (Brenner et al., 1979). Further research must establish whether these antagonistic actions of MPA on peripheral estrogen regulation of olfactory (and/or tactile) factors are responsible for the alterations in sexual attractiveness in stumptailed macaques.

In conclusion, the present studies show that female sexual attractiveness is reduced following

MPA administration, but the magnitude of this effect is dependent upon socio-environmental factors. Among these factors are the amount of social contact available to subjects (i.e., in a social group vs. in time-limited tests) and variations between individuals (e.g., in sexual or social preferences). Finally, a likely mechanism for this reduction in attractiveness is through the antagonistic action of MPA on estrogen regulation of vaginal olfactory and tactile qualities.

ACKNOWLEDGMENTS

We thank the Harry Frank Guggenheim Foundation for its generous support of the research on which this paper is based. We also acknowledge the support of part of this research by USPHS Grant MH 13006-05, and the Rockland Research Institute, Orangeburg, New York. We thank Dr. R. M. Shelden for preparation of the hormone RIA's and his good advice throughout the project. We also thank Ms. A. M. Hogan for her aid during trio testing and blood sample collections, Mr. T. O'Keeffe for his contribution toward establishment of the island colony and development of behavioral observation procedures, Doctors T. Perper and T. Reynolds for aid in data analysis, and Ms. Sylva Grossman for editorial assistance and typing of the manuscript.

REFERENCES

Adams, D.B., Gold, A.R., and Burt, A.D. Rise in female-initiated sexual activity at ovulation and its suppression by oral contraceptives. New Eng. J. Med. 299, 1145-1150 (1978).

Altmann, J. Observational study of behavior: Sampling methods. Behaviour 49, 227-267 (1974).

Bancroft, J. The relationship between hormones and sexual behavior in humans, in Biological Determinants of Sexual Behavior, J.B. Hutchinson, ed. John Wiley, New York (1978), pp. 492-519.

Baum, M.J. Progesterone and sexual attractivity in female primates, in Recent Advances in Primatology, Vol. 1, D.J. Chivers and J. Herbert, eds. Academic Press, New York (1978), pp. 463-474.

Baum, M.J., Keverne, E.B., Everitt, B.J., Herbert, J., and DeGreef, W.J. Effects of progesterone and estradiol on sexual attractivity of female rhesus monkeys. Physiol. Behav. 18, 659-670 (1977).

Baxter, M.J. and Fedigan, L.M. Grooming and consort selection in a troop of Japanese monkeys (Macaca fuscata). Arch. Sex. Behav. 8, 445-458 (1979).

Beach, F.A. Sexual attractivity, proceptivity, and receptivity in female mammals. Horm. Behav. 7, 105-138 (1976).

Bonsall, R.W., Zumpe, D., and Michael, R.P. Menstrual cycle influences on operant behavior of female rhesus monkeys. J. Comp. Physiol. Psychol. 92, 846-855 (1978).

Brenner, R.M., West, N.B., Norman, R.L., and Sandow, B.A. Progesterone suppression of the estradiol receptor in the reproductive tract of macaques, cats, and hamsters, in Steroid Hormone Receptor Systems, W.L. Leavitt and J.H. Clark, eds. Plenum, New York (1979), pp. 173-196.

Chevalier-Skolnikoff, S. Heterosexual copulatory patterns in stumptail macaques (Macaca arctoides) and in other macaque species. Arch. Sex. Behav. 4, 199-220 (1975).

Cochran, C.G. Proceptive patterns of behavior throughout the menstrual cycle in female rhesus monkeys. Behav. Neur. Biol. 27, 342-353 (1979).

Czaja, J.A. and Bielert, C. Female rhesus sexual behavior and distance to a male partner: Relation to stage of the menstrual cycle. Arch. Sex. Behav. 4, 583-597 (1975).

Eaton, G.G. and Resko, J.A. Ovarian hormones and sexual behavior in Macaca nemestrina. J. Comp. and Physiol. Psychol. 86, 919-925 (1974).

Enomoto, T., Seiki, K., and Haruki, Y. On the correlation between sexual behavior and ovarian hormone levels during the menstrual cycle in captive Japanese monkeys. Primates 20, 563-570 (1979).

Esser, A.H., Deutsch, R.D., and Wolff, M. Social behavior adaptation of gibbon (Hylobates lar) in a controlled environment. Primates 20, 95-108 (1979).

Fotherby, K. and Goldzieher, J.W. Animal models for the development of long-acting injectable contraceptives, in <u>Animal Models in Human Reproduction</u>, M. Serio and L. Martini, eds. Raven Press, New York (1980), pp. 461-473.

Goldfoot, D.A. Hormonal and social determinants of sexual behaviour in the pigtail monkey (<u>Macaca nemestrina</u>), in <u>Normal and Abnormal Development of Brain and Behaviour</u>, G.B.A. Stoelinga and J.J. van der Werff ten Bosch, eds. Leiden University Press, Leiden, Holland (1971), pp. 325-341.

Goldfoot, D.A., Slob, A.K., Scheffler, G., Robinson, J.A., Wiegand, S.J., and Cords, J. Multiple ejaculations during prolonged sexual tests and lack of resultant serum testosterone increases in male stumptail macaques (M. arctoides). <u>Arch. Sex. Behav.</u> 4, 547-560 (1975).

Grant, E.G.G. and Mears, E. Mental effects of oral contraceptives. <u>Lancet</u> 2, 945-946 (1967).

Grant, E.G.G. and Pryse-Davies, J. Effect of oral contraception on depressive mood changes and on endometrial monoamine oxidase and phosphatases. <u>Br. Med. J.</u> 3, 777-780 (1968).

Herbert, J. The social modification of sexual and other behavior in the rhesus monkey, in <u>Progress in Primatology</u>, D. Starck, R. Schneider, and J.H. Kuhn, eds. G. Fischer, Stuttgart (1967), pp. 232-246.

Keverne, E.B. Sexual receptivity and attractiveness in the female rhesus monkey, in <u>Advances in the Study of Behavior</u>, Vol. 7, J.S. Rosenblatt, R.A. Hinde, E. Shaw, and C. Beer, eds. Academic Press, New York (1976), pp. 155-200.

Johnson, D.F. and Phoenix, C.H. Sexual behavior and hormone levels during the menstrual cycle of rhesus monkeys. <u>Horm. Behav.</u> 11, 160-174 (1978).

Michael, R.P. and Keverne, E.B. Pheromones in the communication of sexual status in primates. <u>Nature</u> 218, 746-749 (1968).

Michael, R.P. and Bonsall, R.W. Hormones and sexual behavior of rhesus monkeys, in <u>Endocrine Control of Sexual Behavior</u>, C. Beyer, ed. Raven Press, New York (1979), pp. 279-302.

Mora, G. and Johansson, E.D.B. Plasma levels of medroxyprogesterone acetate (MPA), estradiol and progesterone in the rhesus monkey after intramuscular administration of Depo-Provera^R. Contraception 14, 343-350 (1976).

Rall, H.J.S., Ferreira, G.S., and Janseens, K.Y. Effect of medroxyprogesterone acetate contraception on cytoplasmic estrogen receptor content of the human cervix uteri. Int. J. Fertil. 23, 51-56 (1978).

Rosenfield, A.G. Injectable long-acting progestogen contraception: A neglected modality. Am. J. Obstet. Gynecol. 120, 537-549 (1974).

Saayman, G.S. Effects of ovarian hormones on the sexual skin and behavior of ovariectomized baboons (Papio ursinus) under free-ranging conditions, in Symp. IVth Int. Congr. Primat., Vol. 2: Primate Reproductive Behavior. Karger, Basel (1973), pp. 64-98.

Slob, A.K., Wiegand, S.J., Goy, R.W., and Robinson, J.A. Heterosexual interactions in laboratory-housed stumptail macaques (Macaca arctoides): Observations during the menstrual cycle and after ovariectomy. Hormones and Behavior 10, 191-211 (1978a).

Slob, A.K., Baum, M.J., and Schenck, P.E. Effects of the menstrual cycle, social grouping and exogenous progesterone on heterosexual interaction in laboratory housed stumptail macaques (Macaca arctoides). Physiol. and Behav. 21, 915-921 (1978b).

Steklis, H.D., Linn, G.S., Howard, S.M., Kling, A.S., and Tiger, L. Effects of medroxyprogesterone acetate on socio-sexual behavior of stumptail macaques. Physiol. and Behav. 28, 535-544 (1982).

Udry, J.R. and Morris, N.M. Distribution of coitus in the menstrual cycle. Nature 220, 593-596 (1968).

Udry, J.R. and Morris, N.M. The effect of contraceptive pills on the distribution of sexual activity in the menstrual cycle. Nature 227, 502-503 (1970).

Wolfe, L. Behavioral patterns of estrous females of the Arashiyama West troop of Japanese macaques (Macaca fuscata). Primates 20, 525-534 (1979).

Zumpe, D. and Michael, R.P. Ovarian hormones and female sexual invitation in captive rhesus monkeys (Macaca mulatta). Anim. Behav. 18, 293-301 (1970).

Zumpe, D. and Michael, R.P. Effects of ejaculations by males on the sexual invitations of female rhesus monkeys (Macaca mulatta). Behaviour 60, 260-277 (1977).

Addendum to Chapter 5
A Comment on Cross-Specific Comparison of the Effects of Progesterone Treatment on Social Behavior

L. Tiger

There are various elements of the study by Steklis et al. (Chapter 5, this volume), which seem worth speculating on as far as their implications are concerned. In particular it seems worth asking: what is the relevance of the experiment to those concerned with the effects of the administration of contraceptive agents on human females. There has been in my view a remarkable inattention to the social structural impact of such a potent drug-use system as contraceptive pills; that is to say, there may be a failure to realize what the impact is on the macro-structural level of individual responses at the micro-level, such as those reflected in the experiment.

If only on theoretical grounds it should be no surprise that a major change in the reproductive state of significant elements of the breeding pool will cause changes in socio-structural behavior. Put simply, the question arises, does the administration of contraceptive agents to human females produce an impact on their sexual behavior and on the behavior of the males with whom they are in contact? (We assume that emotional and symbolic interactions are behaviors too, and that there may well be an impact of the regimen on these phenomena.) Obviously one impact will be a greater, if not total, uncoupling of intercourse and reproduction, and hence an enhanced attention to intercourse as a feature of personal enjoyment, pleasure seeking, interpersonal exploration and so on. Yet the intensity and complexity of sexual intercourse is related to its role in natural and sexual selection and is obviously to some extent a phylogenetically based phenomenon. This must mean that elements remain of aspects of its role even in acts of intercourse specifically defined as having no reproductive function.

I have elsewhere commented on McClintock's finding that pill use affects the relations between human females (McClintock, 1971; Tiger, 1975) and I have also hypothesized that the regime is with impact on the behavior of human males (Tiger, 1979). Rather than begin with the question: why should introduction of contraceptive chemistry affect reproductive and neo-reproductive behavior, perhaps the appropriate question is why on earth would it not? Insofar as sexual relations reflect the central evolutionary process and insofar as -- as we know from commentators ranging from the students of poetic symbolism to depth psychiatrists -- the role of sexuality is highly ramified in even evidently non-sexual activities, then one must almost expect that, if nothing else, the knowledge that sexuality is de-coupled from reproduction must have effect on its character and on people's expectations from it.

Translating results from experiment with one species to the behavior of another is obviously complex and difficult. There are enormous differences in the species specific sexual behavior of the stumptailed macaque and humans, and even in the regimen our subjects sustained and the twenty-dose a month pattern of human females. Nevertheless, the strong direction of the experimental results and some generalized data about humans point to what might be comparable, or at least analogical effects of the drug regime. Such gross but real phenomena as birth rates considerably below replacement in those industrial societies where pill use is most widespread, the evidently considerable instability of mating pairs reflected in divorce rates among, for example, currently marrying Americans of some forty-five to fifty percent, suggests that changes at the personal and dyadic level may become manifest in much larger scale social events.

It must be considered that never before in mammalian history has the locus of control of reproduction shifted so markedly to one sex. What has always been and remains in other species a matter of the most complicated interactive decision-making between male and female has, at least in potential terms, now become a wholly female-controlled process, thus markedly changing the respective roles in the evolutionary scheme of males and females.

For example, never before has the problem of male paternity certainty been so forcefully exacerbated by the fact that women possess precise and controllable contraceptive methods about which males need have no information. It is well to remember that some fifteen to twenty years ago the principal contraceptive in use was the condom, a socially acknowledgeable device. However, the intrusion of the industrial system into the body through the contraceptive drug or the I.U.D. or sterilization is a new departure in broad biological terms. I must stress that I am in so sense advocating the return to catch-as-catch-can contraceptive methods, nor one which offers females less control over their reproductive destinies than they have currently available to them through the extant array of contraceptive techniques. Nevertheless, the existence of such effective female controlled contraceptives, of which the pill is still the most widely used, cannot but have consequences for males in interaction with females who are in historical terms newly autonomous with respect to decisions about fecundity.

Not only paternity certainty is involved, but there are also possible effects on males and females as far as proceptive, receptive and intercoursing behaviors are concerned. Let us speculate further and remark on the curious historical datum that in the major industrial countries widespread public pressure for liberal abortion laws emerged <u>after</u> the introduction of efficient female-controlled contraceptives. Why should this counter-intuitive situation have developed? One possible reason: insofar as we know, traditionally some one-third to one-half of marriages were contracted during pregnancy and so becoming pregnant was part of courtship and clearly part of the reproductive system and an obligation seemed clear in a relatively large number of cases for males impregnating females to maintain responsibility for this interaction through to child-bearing. However, the loss of male paternity certainty, under the conditions I've already described, have reduced drastically the likelihood that unmarried couples would feel obligated to or desirous of marriage, given the decreased interactivity resulting from increased female autonomy. In other words, males are now able to eschew responsibility for possible progeny, claiming that females' opportunity for successful contraception removes from males the responsibility for its failure or its absence. As Kristin Luker (1975) has shown among women presenting themselves to an abortion clinic in the Berkeley area, only ten percent received support from the men with whom they had been involved, as far as financial assistance for abortion was concerned. Luker also shows that the use and non-use of contraception may be an important element in females' reproductive strategies; misguidedly or not, many women still elect to rely either on the chance of avoiding pregnancy or on the possibility of marrying while pregnant to avoid using -- at least in the relatively sophisticated group Luker interviewed -- relatively easily available and simple-to-use methods of birth control.

Apart from obvious changes in cognitive state which result from contraceptive use, what other mechanisms might change their action? One cannot assume that pheromonal or other interactions in the body depend solely or largely on the actions of contraceptive agents. Nevertheless, insofar as the effect of pheromones was virtually unknown some twenty years ago, the possibility exists that their impact even on relatively complex creatures such as humans living in industrial societies may be greater than anticipated, if only at a relatively low level of intensity. The most important mechanism, however, is presumably the cognitive one. Thought being a sexual behavior too, it is worth asking whether what is a relatively low -- that is, parsimonious -- level of interference among stumptails consequent on pill use may be magnified in the human case because of the effect of thought and cognition in extending and enhancing the meaning and expression of inner states. Does a human female using contraceptive agents change her sexual behavior in a more marked way because she understands the implication of her decision than do other primates, on whose behavior the impact of a chemical agent will be technical and relatively clear?

There are no easy ways to achieve certainty about such assertion. Nevertheless, again on theoretical grounds it would appear hardly strange that a major shift in the locus of control over the central biological interaction could not produce commensurate changes in patterns of social and sexual behavior. It will therefore not be surprising that with quite overwhelming speed females who now have relative autonomy in the reproductive sphere are also achieving relative autonomy in the productive one. Much less often than before will women move from the support of one family funded by a father, uncle, mother, aunt, etc., to another funded by a husband and progenitor. Again, while there is a large distance from the behavior of the monkeys on Hall's Island to the reproductive and sexual experience of a large population in a complex country, nevertheless, there may be reason to consider such broad effects as part of the variables under review when the efficacy and consequence of drug usage is being considered by the pertinent researchers and authorities.

REFERENCES

Luker, K. Taking Chances: Abortion and the Decision Not to Contracept. University of California Press, Berkeley, Calif. (1975).
McClintock, M.K. Menstrual synchrony and suppression. Nature 229, 244-245 (1971).
Tiger, L. Somatic factors and social behavior, in Biosocial Anthropology, R. Fox, ed. Malaby Press, London (1975).
Tiger, L. Optimism: The Biology of Hope. Simon and Schuster, New York (1979).

Chapter 6
Social and Sexual Behavior During the Menstrual Cycle in a Colony of Stumptail Macaques *(Macaca Arctoides)*

N.C. Harvey

INTRODUCTION

Objectives

At present no information is available on the sexual behavior during the menstrual cycle of stumptail macaques (Macaca arctoides) constantly maintained in an established social group. Studies of sexual behavior of stumptail macaques under laboratory testing conditions (where males and females are housed separately until a timed sex test) have indicated that social and sexual interactions between the pairs remain relatively unaffected by the presence of ovarian and adrenal hormones (Slob et al., 1978a,b; Goldfoot et al., 1978; and Baum et al., 1978).

These findings from laboratory experiments on stumptails are quite different from results of similar experiments which have been reported for the rhesus macaques (Everitt et al., 1978; Michael et al., 1972; Dixson et al., 1973; Johnson and Phoenix, 1976; and Wallen and Goy, 1977) and for pigtail macaques (Goldfoot, 1971; Eaton and Resko, 1974) where both ovarian and adrenal hormones have been implicated in the regulation of sexual behavior.

It has been suggested that social and environmental factors such as female rank, sexual preferences between individual pairs and the constant separation and reunion of the opposite sexed pairs may have masked the influence of hormonal factors on sexual behavior in the stumptail macaques (Goldfoot et al., 1978; Slob et al., 1978b).

The primary objectives of this investigation were to examine social and sexual behavioral interaction throughout the menstrual cycle in colony living stumptail macaques in order to determine whether under these environmental conditions behavioral interactions between members are affected by the presumed underlying hormonal condition of the cycling female. Social factors, such as the dominant and subordinate relationships between colony members, kinship relationships between males and females and individual preferences between the sexes were taken into consideration in the analyses of social and sexual behavior during the menstrual cycle.

The comparative study of social and sexual behavior in the non-human primate is of interest to many scientists concerned with evolutionary biology. Such studies are required for the construction of behavioral models for the underlying mechanisms involved in human behavior.

METHODS

Subjects

The stumptail colony was established in the spring of 1968 at the University of California, Riverside (UCR), under the supervision of Ramon Rhine (Rhine and Kronenwetter, 1972; Rhine, 1972). The animals were imported from Thailand and were of reproductive age when they arrived. At the start of the study, the social group contained eight adult females, seven adult males and three infants. Two more infants were born during the study period. The colony consisted of six matrifocal units. All the animals with the exception of one adult male belonged to a matrifocal unit. Six of the adult males and two of the adult females were colony born and were 6-7 years of age. Exact ages of the imported animals were unknown, but it was estimated that the six imported females were at least 12 years old and the male about 14 years old at the time of this study.

Housing and Maintenance

The animals were housed in an outside enclosure which consisted of two sections, each 20 feet long, 12 feet deep and 7 feet high. The animals were maintained on monkey chow. Additional foods, including fresh fruits, vegetables, wheat and rice, were provided two times a week after the vaginal swabbing of females was completed.

Behavioral Sampling

Two methods of behavioral sampling were utilized for this study: a one-zero focal sampling technique and a specific behavioral sampling system.

Each adult female was subject to a ten-minute focal sampling period each observation day. A one-zero frequency behavioral check list containing active and passive categories was used to record social interactions. A behavior was scored once if it occured one or more times during a 30 second interval.

Copulatory and aggressive behaviors occurred less often than social affiliative behaviors in the captive colony, and rarely occurred during focal sampling. Therefore, these behaviors were recorded whenever they occurred. This was possible because observation conditions were excellent and the behaviors were sufficiently "attention-attracting" (Altmann, 1974) to allow instances to be recorded.

Behavioral Definitions

Table 1 lists the behaviors and definitions employed during the study period. They were adapted and enlarged from definitions previously used by Rhine (1972). Measures of proceptivity included approaching males, proximity to males and spontaneous presents. Measures of attractiveness included approaching females, contacting or threatening females to present, sex investigation, incomplete and complete copulation. While observational conditions and sampling procedures permitted recording of copulations, they were not always conducive to uniform recording of the initial behaviors which ultimately led to copulation. Therefore, female receptivity as measured by the acceptance ratio of the number of female presents in response to male contact was not employed for this study.

Detection of Menstruation

Starting in December, 1976, the stumptail females were trained to accept biweekly vaginal swabs within the confines of the enclosure. In addition to vaginal swabbing, female perinea were visually inspected every observation day for external signs of menses. Although Macaca arctoides is not a species which typically exhibits external menstruation, there were four individuals in the colony who did.

Observation Schedule

There were 268 focal observations days scheduled for five days a week, Monday through Friday between October 18, 1976, and October 28, 1977. There were 66 missing scheduled focal observation days due to holiday breaks, inclement weather and illnesses on the part of the observer. Therefore, there were a total of 505 focal hours of observation based on 202 observation days where each of the 15 adult animals was subjected to a 10 minute focal observation period. There were 303 hours spent sampling specific behaviors as they occurred during the 202 observation days. In addition to the 202 days where female perinea were examined for external signs of menstruation, there were an additional 16 days where this procedure was maintained in spite of the inclement weather. Focal sampling and sampling of specific behaviors were conducted throughout the day from 0730 to 1700 hours.

Table 1. Behavioral Categories and Definitions

BEHAVIOR	DEFINITIONS
Approach	Approaching within 2 feet of another in a purposeful manner, usually with eyes directed at the other.
Proximity	Remaining within 2 feet of another, but <u>not</u> touching.
Contact	Any body contact between individuals (e.g., extensive huddle while resting, a hand on another, back to back sitting); does <u>not</u> include grooming, copulation or aggression.
Groom	Examination of the hair, skin, and body wounds of another individual by hand and occasionally by mouth.
Present	Turns body sidewards or orients hindquarters toward another, frequently accompanied by individual turning to look at presenter.
Sex Investigate	Individual inspects perineum area of another by visual, olfactory and gustatory examination of the vaginal tract.
Incomplete Copulation	Hands on hip of another, feet grasp ankles of other, accompanied by thrusting, <u>ejaculation does not occur</u>, and an erection is visible after the mount.
Complete Copulation	Same as above, but ejaculation occurs. Easily recognizable with male's body rigid, open mouth, male pulls female into lap for "pair-sit" and ejaculate visible on male.
Threat	Includes a fixed stare to slightly open mouth, full open mouth, eyebrow raise, lowered head to a short lunge. It also includes slap, hit, mouth bite, shove and display-hop (branch-shake).
Avoid	Where an individual actively maintains a distance of two feet of more from another who has not been paying any sort of attention to that individual.
Displace	Occurs when an individual actually takes over a resting spot, preferred food, or grooming partner, or sexual partner of another.

Data Analyses

Hormonal evaluation during the menstrual cycle for Macaca arctoides indicated that endocrinologically normal cycles ranged between 26 to 32 days in duration (Wilks, 1977; Slob et al., 1978a). Brügenmann and Grauwiler (1972) showed cycles ranged between 26 and 40 days in duration (with data on 42 stumptails for a total of 319 cycles).

Because of the above reports, the decision for this study was made to consider cycles as normal if they ranged between 26 and 40 days.

In all, 34 cycles from 7 colony females were identified and used in behavioral analyses for this study. The cycle lengths ranged from 26 to 32 days, with a mean of 28.6 days ($\pm$ 1.90 SD).

In the absence of hormone assay information, the time of ovulation could not be determined for the UCR colony females. In addition, because five of the seven females became pregnant during the course of the study, the menstrual cycle days could not be counted backwards from the first day of menses, a procedure which reduces the variability in the duration of the follicular phase (Michael and Zumpe, 1970). Therefore, the menstrual cycles were counted forward from day 1 of menstruation and midcycle was defined as days 11-15 of the cycle, a procedure used in stumptail breeding by Macdonald (1971). For this study the menstrual cycle was divided into three phases: (1) follicular, (2) midcycle, and (3) luteal.

A mean frequency score for each behavior for the number of days in each cycle phase was calculated from each focal animal's interaction with other colony members. For each category, the mean scores obtained for each cycle phase were compared using a one way analysis of variance for repeated measures. When significant variation between cycle phases for a behavior was observed, a Newman-Keuls test was employed to determine which comparisons among means of cycle phase were significant. Student's t test for dependent groups was employed to assess effects of rank on behavior.

RESULTS

Non-Sexual Behavior

The dominance hierarchy for the stumptail colony was based on the combined frequencies of displace, threat, and avoid. Mean scores among all animals were tabulated in a 15 x 15 matrix. Since many dyads in the colony displayed both dominant and submissive behaviors to each other, scores for each dyad were subtracted to determine the dominant animals per dyad. Resulting means in a half matrix revealed the dominance structure of the social group.

Table 2 shows both the dominance hierarchy and kinship of the adult animals in the colony. As shown, only two of the adult males, Paul and Ute, were dominant over all the adult females in the colony. Ute was one of the original breeding males and the former alpha male prior to the start of this study. During this investigation, Ute was still dominant over Maria and Emma. However, both of these females were related to the dominant male Paul. Therefore, Paul would frequently protect and support his related females from agonistic encounters with Ute. The general impression from observations was that Ute's dominance over Maria and Emma was tenuous and may have been diminishing during the course of this study.

There were only two exceptions where colony born males were dominant over their mothers. Paul, the alpha male, was dominant over his mother Maria, and Sam, the third ranking male, was dominant over his mother Gail. The remaining colony-born males were subordinate to their mothers.

There were two brother-sister pairs in the colony. In one pair, Paul was dominant over his sister Emma, while in the other pair, Ned was subordinate to his sister Queen.

Females varied significantly in their approach of males ($P < .05$) during the menstrual cycle, with mean values being greater around midcycle ($P < .05$) than during the follicular phase. Grooming of males by females also varied significantly ($P < .05$) among cycle phases, with higher mean values

Table 2. Dominance hierarchy in group living Macaca arctoides as measured by the combined frequencies of displacement, threat and avoid receive. The dominant animal is represented by the row of the cell and subordinate animal by the column of the cell.

	Paul	Ute	Maria	Emma	Sam	Gail	Heather	Ivan	Joan	Thelma	Queen	Vic	Ned	Lois	Russ
Paul	-	.157	.233*	.316*	.110	.333	.227	.200	.175	.243	.285	.223	.200	.289	.354
Ute	0	-	.010	.009	.047	.121	.106	.075	.103	.081	.083	.060	.062	.143	.049
Maria	0	0	-	.161*	.029	.122	.122	.067	.066	.090	.064	.134	.076	.049	.043
Emma	0	0	0	-	.067	.055	.177	.070	.126	.147	.110	.167	.172	.131	.121
Sam	0	0	0	0	-	.016*	.023	.042	.064	.016	.058	.080	.051	.029	.093
Gail	0	0	0	0	0	-	.181	.034	.030	.076	.049	.023	.038	.065	.013
Heather	0	0	0	0	0	0	-	.017*	.063	.072	.076	.033	.031	.058	.024
Ivan	0	0	0	0	0	0	0	-	.035	.016	.044	.060	.070	.010	.045
Joan	0	0	0	0	0	0	0	0	-	.073	.099*	.035	.083*	.064	.025
Thelma	0	0	0	0	0	0	0	0	0	-	.015	.030*	.024	.043	.020
Queen	0	0	0	0	0	0	0	0	0	0	-	.020	.027*	.009	.010
Vic	0	0	0	0	0	0	0	0	0	0	0	-	.053	.010	.112
Ned	0	0	0	0	0	0	0	0	0	0	0	0	-	.035	.035
Lois	0	0	0	0	0	0	0	0	0	0	0	0	0	-	.031*
Russ	0	0	0	0	0	0	0	0	0	0	0	0	0	0	-

* Asterisked values represent kinship between column and row animals.

occurring around midcycle ($P<.05$) than in either the follicular or luteal phases. Time spent in proximity to males was greatest at midcycle; however, analysis of variance did not reveal any signficiant changes over cycle phases. While stumptail females did not show any variation during the cycle in displacing or avoiding males, they did exhibit significant variation in threatening males ($P<.05$), with significantly higher mean values occurring during the follicular phase ($P<.05$) than at midcycle (Table 3).

Colony males did not alter their approach behavior toward cycling females. Instead, males were observed to approach females at a fairly constant rate throughout the cycle. Males varied their grooming of females significantly ($P<.05$) over the three phases of the menstrual cycle, with values being higher at midcycle than in the luteal phase, and this difference was significant ($P<.05$). Females were not threatened by or avoided by males any differently among cycle phases. There was a slight decrease in male displacement of females during the follicular phase, but again this change was not significant (Table 3).

Since females displayed significant variation in their behavior with males, additional analyses of variance were undertaken to determine if the stumptail females varied their behavior over the course of the menstrual cycle with their kin-related males. It was found that the females did not significantly vary in either their approach, grooming or threat behavior towards their kin-related males. In turn, kin males did not vary in numbers of approaches or in time spent grooming their related females during the menstrual cycle. Although males displayed higher mean values in threatening their related females during the follicular phase of the cycle, the difference was not significant. Kin-related animals in the colony rarely avoided or displaced each other, therefore the extremely small mean values found between phases were not subject to analyses of variance (Table 4).

While females showed no significant changes in their affiliative behavior towards their related males, they did exhibit significant variation in

Table 3. Various behaviors by cycling females to males, and males' behavior to cycling females during follicular, midcycle and luteal phases of the menstrual cycle in colony living stumptail macaques (Macaca arctoides).

Behavior	N[1]	Follicular 1-10 days Mean and SEM	Midcycle 11-15 days Mean and SEM	Luteal 16-28 days Mean and SEM	F[2]	P
To males						
Approach	7	.56 ± .12	.95 ± .12[b]	.68 ± .14	6.86	<.05
Groom	7	1.74 ± .51	2.57 ± .57[a]	1.27 ± .49	5.22	<.05
Proximity	7	2.96 ± .20	3.81 ± .73	2.49 ± .39	2.20	
Threat	7	.22 ± .05[c]	.07 ± .03	.10 ± .04	5.18	<.05
Displace	7	.09 ± .05	.07 ± .04	.09 ± .04	1.49	
Avoid	7	.33 ± .09	.43 ± .14	.46 ± .12	.59	
By males						
Approached	7	1.19 ± .20	1.37 ± .25	.96 ± .20	1.39	
Groomed	7	1.46 ± .42	3.21 ± .93[d]	.95 ± .17	5.33	<.05
Threatened	7	.31 ± .06	.38 ± .11	.23 ± .08	.99	
Displaced	7	.11 ± .03	.24 ± .05	.19 ± .06	2.27	
Avoided	7	.04 ± .03	.08 ± .04	.05 ± .03	.56	

1. Number of females
2. One way analysis of variance, with 2, 12 df for all F values.

Newman-Keuls, P <.05: a significantly different from follicular and luteal phases.
b significantly different from follicular phase.
c significantly different from midcycle phase.
d significantly different from luteal phase.

Table 4. Various behaviors by cycling females to kin-related males, and kin males' behavior to cycling females during follicular, midcycle and luteal phases of the menstrual cycle in colony living stumptail macaques (Macaca arctoides).

Behavior	N[1]	Follicular 1-10 days Mean and SEM	Midcycle 11-15 days Mean and SEM	Luteal 16-28 days Mean and SEM	F[2]	p
To kin males						
Approach	7	.13 ±.03	.13 ±.04	.09 ±.03	.71	
Groom	7	.49 ±.21	.54 ±.25	.26 ±.15	.91	
Proximity	7	1.11 ±.26	.68 ±.25	.43 ±.15	3.16	
Threat	7	.06 ±.03	.02 ±.02	.02 ±.01	.81	
By Kin males						
Approached	7	.27 ±.07	.22 ±.07	.25 ±.09	.21	
Groomed	7	.30 ±.12	.51 ±.24	.18 ±.06	1.35	
Threatened	7	.14 ±.05[a]	.04 ±.03	.03 ±.02	3.86	

1. Number of females
2. One way analysis of variance, with 2, 12 df for all F values.

their approach to the non-related males in the colony ($P<.01$), with higher mean levels occurring during midcycle than in the follicular phase ($P<.05$). Although females showed an increase during midcycle in both grooming and being proximate to non-related males, neither of these behaviors varied significantly. Colony females exhibited significant variation in threatening non-related males ($P<.01$) with higher mean values occurring during midcycle and luteal phases, and these differences between means were significant ($P<.05$). The females did not show any changes in either displacing or avoiding non-related males between cycle phases (Table 5).

The non-related males varied significantly in their grooming of the colony females ($P<.05$) during the menstrual cycle. Comparison between means revealed a significant difference between midcycle and luteal phases ($P<.05$). Stumptail females were not approached, threatened, avoided or displaced any differently among cycle phases by the non-related males in the colony (Table 5).

Sexual Behavior

Since presenting behavior within a group serves a social function as well as a sexual one, it was relevant to initially consider female presenting as a social act.

Table 6 shows the distribution of means for all female presents directed towards individual males, regardless of cycle phase. Males were presented to by females according to the male's social rank, with the highest ranking male receiving the greater mean number of presents (Spearman's $\underline{rho}$-0.938, N=7, $P<.01$). A similar pattern was $\underline{not}$ observed for female rank and presenting behavior. Instead, female presenting was a function of the female's rank position to the male. A $\underline{t}$ test revealed females presented significantly more (t=5.51, 6df, $P<.001$) to males higher ranking than themselves than to males who were lower ranking than they were. Therefore, in a social group, presenting behavior on the part of the stumptail females is a function of

Table 5. Various behaviors by cycling females to non-related males, and non-related males' behavior to cycling females during follicular, midcycle and luteal phases of the menstrual cycle in colony living stumptail macaques (Macaca arctoides).

Behavior	N[1]	Follicular 1-10 days Mean and SEM	Midcycle 11-15 days Mean and SEM	Luteal 16-28 days Mean and SEM	F[2]	P
To non-related males						
Approach	7	.42 ± .10	.80 ± .11[a]	.59 ± .14	8.43	<.01
Groom	7	1.24 ± .45	2.04 ± .73	1.01 ± .42	2.54	
Proximity	7	1.85 ± .25	3.13 ± .80	2.06 ± .42	1.99	
Threat	7	.16 ± .04[b]	.05 ± .03	.08 ± .04	8.07	<.01
Displace	7	.09 ± .05	.07 ± .04	.08 ± .04	1.35	
Avoid	7	.30 ± .10	.37 ± .16	.39 ± .14	.30	
By non-related males						
Approached	7	.91 ± .24	1.15 ± .27	.70 ± .21	2.03	
Groomed	7	1.15 ± .39	2.70 ± .99[c]	.76 ± .16	3.70	<.06
Threatened	7	.16 ± .04	.34 ± .13	.19 ± .08	1.58	
Displaced	7	.08 ± .04	.16 ± .06	.15 ± .07	1.01	
Avoided	7	.04 ± .02	.05 ± .00	.04 ± .02	.07	

1. Number of females
2. One way analysis of variance, with 2, 12 df for all F values.

Newman-Keuls, $P < .05$:
 [a] significantly different from follicular phase.
 [b] significantly different from midcycle and luteal phases.
 [c] significantly different from luteal phase.

the dominant and subordinate relationships between males and females. Within these social parameters, female presenting behavior will be examined in relation to the phases of the menstrual cycle.

Presents were categorized as either spontaneous or induced. Spontaneous presents consisted of: (1) female approach and present and (2) female proximity and present. Induced presents were a response to (1) male approach and (2) male contact or threat.

Female <u>approach and present</u> varied over the menstrual cycle ($P<.06$), with significantly higher mean values occurring at midcycle than in the luteal phase ($P<.05$). However, female <u>proximity and present</u> did not change with phases of the cycle (Table 7).

Only the two highest ranking females (Maria and Emma) were observed approaching and presenting to their related male during the menstrual cycle. Both females exhibited a decrease in their approach and present to their kin male at midcycle. More importantly, at midcycle these females were not observed approaching and presenting to the five males lower ranking than themselves. When data on these two females were excluded, there were significant changes over the phases of the menstrual cycle ($F=9.82$, 2,10df, $P<.01$) for approach and present to colony males, with the increase at midcycle being significantly different from the luteal phase ($P<.05$).

Female presents induced by <u>male approach</u> showed no variation over cycle phases. Presents induced by male <u>contact or threat</u> did vary among cycle phases ($P<.05$), with males contacting or threatening females to present significantly more at midcycle than during the luteal phases of the cycle ($P<.05$), as shown in Table 7.

Females were seldom contacted or threatened by their kin-related males during the menstrual cycle. Only three females (Maria, Emma and Gail) were induced to present in this manner. In each case, the female was subordinate to her related male.

When all presents were combined together for the non-related males in the colony, there was a significant phase effect ($P<.05$), with midcycle

Table 6. Mean number of female presents to colony males independent of cycle phase. Mean values above the dark line represent female dominance over the male. Mean values below the line represent female subordinacy to the male.

Females In Rank Order	Males in Rank Order								$\overline{X}$	SEM
	Paul	Ute	Sam	Ivan	Vic	Ned	Russ	N		
Maria	.68	.08	.01	.01	0	.01	0	7	.11	.09
Emma	.82	.16	.01	.10	0	.01	0	7	.16	.11
Gail	1.93	.55	.05	0	0	0	0	7	.36	.27
Heather	.57	.24	.10	.06	0	.03	0	7	.14	.07
Joan	.78	.62	.43	.03	.03	.05	0	7	.28	.11
Queen	.59	.20	.28	.08	0	0	0	7	.16	.08
Lois	.50	.31	.37	.06	.06	.28	0	7	.22	.07
N	7	7	7	7	7	7	7			
$\overline{X}$	.84	.31	.18	.05	.01	.05	0			
SEM	.18	.07	.07	.01	.01	.04	0			

Table 7. Female presenting behavior and sex investigation by males during the follicular, midcycle and luteal phases of the menstrual cycle in colony living stumptail macaques (Macaca arctoides).

Behavior	N^1	Follicular 1-10 days Mean and SEM	Midcycle 11-15 days Mean and SEM	Luteal 16-28 days Mean and SEM	F^2	P
Spontaneous Presents						
Approach and Present	7	$.49 \pm .05$	$.63 \pm .14^a$	$.34 \pm .107$	3.82	<.06
Prox and Present	7	$.28 \pm .06$	$.22 \pm .07$	$.18 \pm .04$	.78	
Induced Presents						
Male Approach	7	$.25 \pm .09$	$.35 \pm .11$	$.25 \pm .05$	.65	
Male Contact/Threat	7	$.27 \pm .14$	$1.18 \pm .50^b$	$.22 \pm .14$	4.04	<.05
All Presents Combined						
To kin males	7	$.26 \pm .14$	$.22 \pm .16$	$.22 \pm .12$		
To non-kin males	7	$1.02 \pm .29$	$2.16 \pm .68^c$	$.77 \pm .24$	4.65	<.05
Sex Investigation						
By all males	7	$.62 \pm .11$	$.96 \pm .37$	$.45 \pm .11$	1.96	
By kin males	7	$.08 \pm .03$	$.10 \pm .06$	$.10 \pm .05$	.08	
By non-kin males	7	$.54 \pm .14$	$.86 \pm .32$	$.35 \pm .13$	1.91	

1. Number of females.
2. One way analysis of variance with 2, 12 df for all F values.

Newman-Keuls P <.05: a significantly different from luteal phase.
                                    b significantly different from luteal phase.
                                    c significantly different from luteal and follicular phases.

being significantly different from the follicular and luteal phases (P<.05). There was no variation among cycle phases for the combined presents directed toward the kin males in the colony (Table 7).

In addition to the effects of kinship and dominance relationship between males and females, there were other factors which influenced presenting behavior. Some females were considered "more attractive" than others, as measured by males contacting or threatening the females to present. Gail was the only female who was frequently contacted or threatened to present for all three phases of the menstrual cycle. At midcycle, the males found Gail, Joan and Lois all highly attractive while Heather and Queen were rarely contacted or threatened to present and therefore evidently considered unattractive.

Males were most frequently observed sex investigating females during the follicular and midcycle phases, followed by a depression in the luteal phase; however, these changes were highly variable and were not significantly different as a function of cycle phase. In addition, separate analyses of variance for kin and non-kin males did not reveal any significant changes during the menstrual cycle in sex investigation, as shown in Table 7.

As in the case of presenting behavior, male sex investigation was affected by the rank relationship between the male and female. Females were investigated significantly more by males higher ranking than themselves than by lower ranking males (t=3.61, 6df, P<.02).

The same three females (Gail, Joan and Lois) that had high mean values of being contacted or threatened to present at midcycle were also sex investigated more than the other females in the colony. There was no apparent correlation between female rank among all seven females and the frequency of sex investigation by males.

There were a total of 137 copulatory events observed during the 34 menstrual cycles of the seven females: 76 were complete copulations and 61 were incomplete. Females who did not experience complete

Table 8. Number of Menstrual Cycles and Copulatory Events Observed for Each Stumptail Female.

Female	Cycles	Complete	Incomplete	Total
Maria	12.5	0	6	6
Emma	5.0	4	8	12
Gail	4.5	48	23	71
Heather	3.5	2	2	4
Joan	3.5	16	6	22
Queen	3.5	5	6	11
Lois	1.5	1	10	11
Total	34.0	76	61	137

copulations during the menstrual cycle were still included in the data analyses.

Copulatory activity in the UCR colony was relatively infrequent and varied considerably among the females. Table 8 shows the number of cycles and copulatory events observed for each female. Although the variance in numbers of copulations due to individual differences was great, the distribution of copulations in relation to the phase of the menstrual cycle formed a meaningful pattern.

Figure 1 shows the mean number of all copulatory events (both complete and incomplete copulations) for each day of the menstrual cycle. Copulatory behavior occurred throughout the cycle with a distinct peak at midcycle, followed by a second peak in the late luteal phase, a statistically significant ($P < .01$) phase effect (Table 9), with midcycle values being significantly different ($P < .05$) from follicular and luteal phases.

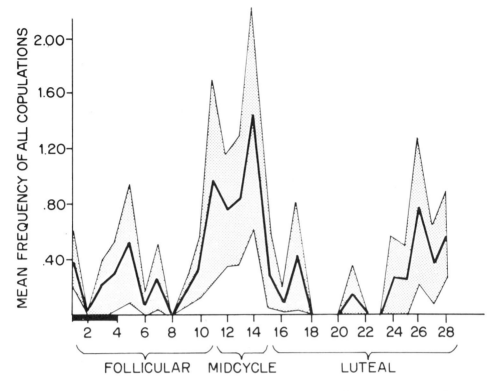

Figure 1. Mean distribution (dark line) and standard error of the mean (shaded area) of all copulations during the menstrual cycle in colony living stumptail macaques (<u>Macaca arctoides</u>). Dark bar indicates menses.

When copulatory events were distinguished as complete and incomplete copulations, it appeared that both types occurred sporadically from days 1-10 and 24-27. There was a distinct peak in numbers of completed and incomplete copulations on midcycle days 11-14. During the luteal phase completed copulations were observed on days 17 and 21, while incomplete ones never occurred from days 17-25.

One female, Gail, received most of the copulations. With exclusion of her data a significant ($P < .05$) effect of cycle phase was still observed (Table 9), with a midcycle peak followed by a second peak in the late luteal phase (Figure 2). The number of completed copulations was greatest at midcycle and perimenstrual portions of the cycle, but without Gail's data the phase effect was not

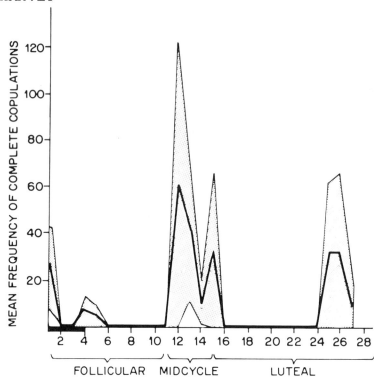

Figure 2. Mean distribution (dark line) and standard error of the mean (shaded area) of complete copulations during the menstrual cycle for six colony females over 29.5 menstrual cycles. Dark bar indicates menses.

statistically significant (Table 8). Nevertheless, with or without Gail's data copulatory activity in colony living stumptail macaques was a function of cycle phase (Figures 1, 2).

Social factors such as rank, kinship, male preference, female attractiveness and male-male competition all contributed to the individual differences observed in numbers of copulations.

In the male, sexual behavior was clearly correlated with his rank: the dominant male (Paul) performed 75% and the beta male (Ute) 23% of the total number of copulations.

Similarly, kinship affected copulatory behavior in this stumptail colony. Overall, complete copulations occurred significantly more than expected with non-related males (P-.05)

Table 9. Copulatory behavior during follicular, midcycle and luteal phases of the menstrual cycle in colony living stumptail macaques (Macaca arctoides).

Behavior	N[1]	Follicular 1-10 days Mean and SEM	Midcycle 11-15 days Mean and SEM	Luteal 16-28 days Mean and SEM	F[2]	P
Incomplete Copulations	7	.05 ± .03	.17 ± .10[a]	.08 ± .05	4.58	<.05
Complete Copulations	7	.17 ± .10	.54 ± .28	.14 ± .10	3.48	<.06
Combined Copulations	7	.21 ± .09	.97 ± .33[b]	.21 ± .15	8.50	<.01
Combined Copulations without Gail	6	.12 ± .03	.75 ± .28[c]	.07 ± .04	5.64	<.05
Complete Copulations without Gail	6	.03 ± .01	.33 ± .22	.04 ± .02	1.86	

1. Number of females
2. One way analysis of variance, with 2, 12 df for 7 N and 2, 10 df for 6 N for all F values.

Newman-Keuls P<.05: [a] significantly different from follicular and luteal phases.
                   [b] significantly different from follicular and luteal phases.
                   [c] significantly different from follicular and luteal phases.

regardless of phase of cycle. At midcycle, differences from the expected for complete copulations with non-related individuals were significant ($P<.02$). Incomplete copulations between kin-related animals for the entire menstrual cycle was more frequent than expected ($P<.10$), but at midcycle there were no significant differences from the expected between kin and non-kin males for incomplete copulations (Table 10). The dominant male, Paul, performed all his kin-related completed copulations with his sister (Emma) and performed incompleted copulations with both his sister and his mother (Maria). The completed copulations with his sister were restricted to the late luteal and early follicular phases of the menstrual cycle.

Gail was not only the most sexually active female in the colony but was the preferred female of the dominant male as indicated by their level of copulatory activity for the menstrual cycle. This preference of Paul for Gail accounted for most of the incomplete and all of the complete copulations observed on days 17 and 21 during the luteal phase of the cycle (Figure 1). The copulations which occurred at this time appeared to be a result of the dominant male, Paul, re-directing to Gail his sexual arousal from his mother when she was at day 10 and 11 of her cycle and from Joan (non-kin) during the early part of her pregnancy. Thus, Paul frequently exhibited high levels of sexual excitement with lipsmacking and partial mounts when he sex investigated these two females, then stopped and threatened Gail to present for sex investigation. However, in each case he appeared frustrated by Gail's unattractive reproductive condition, as he would hit, push and bite her after examining her vagina, and then return to his mother or Joan and sex investigate them again. This switching back and forth between Gail and the other females was repeated 3-7 times before Paul mounted and ejaculated with Gail. Paul's preference for Gail was probably due to the fact that she was the highest ranking, non-related female in the colony.

Interference by the dominant male accounted for 26% of all the incomplete copulations observed

Table 10. Distribution of copulations between kin and non-kin animals during the different phases of the menstrual cycle in colony living stumptail macaques (Macaca arctoides).

	Follicular 1-10 days		Midcycle 11-15 days		Luteal 16-28 days		TOTAL	
	Kin	Non-Kin	Kin	Non-Kin	Kin	Non-Kin	Kin	Non-Kin
Copulation								
Complete	1	12	0	43^2	3	17	4	72^1
Incomplete	4	16	5	22^4	5	9	14^3	47
Total	5	28	5	65	8	26		137

1. Complete copulation for kin versus non-kin, no phase, $X^2 = 4.34$, $P < .05$
2. Complete copulation for kin versus non-kin, midcycle, $X^2 = 6.04$, $P < .02$
3. Incomplete copulation for kin versus non-kin, no phase, $X^2 = 3.07$, $P < .10$
4. Incomplete copulation for kin versus non-kin, midcycle, $X^2 = .13$, N.S.

during the menstrual cycle. Paul was observed attacking, biting and pulling Ute (beta male) off the three most attractive females (Gail, Joan and Lois) at midcycle. Paul's interference accounted for 85% of the incomplete copulations with Lois. Males displayed very little sexual interest in Heather during the menstrual cycle and no competition was observed between Paul and Ute for sexual access to either Heather or Queen at midcycle.

DISCUSSION

These data on changes in social and sexual behavior during the menstrual cycle differ from those previously reported for stumptails studied under laboratory pair-testing conditions and in small temporary groups.

In previous studies none of the reported female behaviors, including approach to male, grooming of male and spontaneous presents were found to vary with cycle phase (Slob et al., 1978a,b). However, the results of the present study are similar to those reported for female pigtail macaques in small groups (Goldfoot, 1971). Cyclical changes in stumptail female behavior are also comparable to changes reported for the rhesus females during the menstrual cycle, even though the particular behaviors involved are somewhat different for the two species (Czaja and Bielert, 1975; Michael and Bonsall, 1977).

The cyclic changes of the males' grooming, threat and copulatory behaviors were different from those previously reported for the stumptail male under laboratory conditions (Slob et al., 1978a,b). However, this study confirms previous findings (Slob et al., 1978a,b) on stumptail macaques: males did not vary in their approach, proximity and sex investigation (genital inspection) of the females during the menstrual cycle, but did vary in inducing the females to present by contact or threat.

The present results on complete copulations during the menstrual cycle are similar to those reported for pigtail macaques (Goldfoot, 1971) in small groups and in a pair-test (Bullock et al.,

1972; Eaton and Resko, 1974), and for rhesus males (Czaja and Bielert, 1975; Michael and Bonsall, 1977).

While receptivity, as measured by the acceptance ratio of the number of female presents in response to male contact, was not included in the present study, it is not clear that it would have shown cyclical variation. The general impression was that stumptail females in this colony usually responded to male contact or threat with a negative adjustment. Furthermore, if the male persisted, there was very little the female could do within the confines of the enclosure. In many cases the male would pursue, corner and attack the female repeatedly until he was successful in coercing her to present and accept his mount. Interestingly, once the male intromitted, the female (who moments before had been resisting the copulation) would display facilitory or stimulatory behaviors, such as looking back at the male with a round-mouth facial expression and hand contacting the male.

Other studies on stumptail (Slob et al., 1978b), pigtail (Goldfoot, 1971) and rhesus (Perachio et al., 1973) macaques have found that female attractiveness is correlated with her social rank. The results of the present study also show that the most dominant male preferred a high ranking female as measured by the frequency of coital activity throughout the menstrual cycle. And yet, female attractiveness was not solely a function of the female's rank, as there was a great deal of male-male competition in copulation at midcycle with the fifth and eighth ranking females but not with the other females in the colony. Therefore, when stumptail macaques are maintained in a multi-male environment, some other factor(s) apparently play a role in female attractiveness.

The differences found between the results of this study and those reported by Slob et al. (1978a) on sexual behavior in the stumptail may be a result of the way the cycles were plotted in each study. In this study the cycles were plotted forward from day one of menses, whereas Slob et al. (1978a) plotted their cycles backwards from day one of menses. In order to determine if this could account for the differences in results, the cycles in which females did not become pregnant (10 cycles for 5 females) were plotted and sampled in the same manner as reported by Slob et al. (1978a).

A single behavior, female spontaneous <u>approach and present</u>, which was found to vary significantly in this study, was re-analyzed. Again significant variation was found ($F=4.98$, 2, 8df, $P<.05$), with females displaying significantly more spontaneous <u>approach and presents</u> around midcycle than during the early follicular or mid-luteal phases ($P<.05$).

Therefore, it is unlikely that the differences between the present results and those of Slob et al. (1978a) are due to the alternative method employed in calculating menstrual cycle phases.

The relevant question at this point is why do stumptails vary their social and sexual behavior during the menstrual cycle in a colony environment and not in a laboratory testing environment? Factors related to this question are considered below.

Stumptail males are significantly larger and heavier than females (Harvey et al., 1979). Furthermore, under pair-testing conditions males are probably subject to increased arousal (perhaps from previous experience with the testing environment). There have been several reports on stumptails describing excessive male aggression towards the female in such environmental conditions (Macdonald, 1971; Trollope and Blurton Jones, 1975; Slob et al., 1979). Therefore, when a female is placed with an already aroused, much larger male in an environment which offers no escape, she probably has very little opportunity to display any variation in her behavior. Goldfoot (1977) has noted that if a primate male is persistent, "rape" can occur in a testing environment. This has been a problem for rhesus macaques in a pair-test situation, according to Michael and Bonsall (1977); the males are larger than the females and male threats and aggression toward the female "contaminate" measures of female behavior. A similar conclusion was reached by Wallen and Goy (1977) concerning measures of receptivity for the rhesus in a pair-test environment.

While this argument may partially explain the inability to find clear evidence of hormonal regulation of proceptivity and receptivity in the stumptail female, it does not account for the continual sexual attractiveness of the female as measured by the male stumptail copulatory performance.

Studies to date clearly indicate that stumptail males continue to ejaculate regardless of the steroid condition of the female (Slob et al., 1978a,b). This is especially evident when females were ovariectomized and/or adrenalectomized (Goldfoot et al., 1978; Baum et al., 1978). In the rhesus, male ejaculation is severely depressed or totally eliminated after ovariectomy and/or adrenalectomy (Goldfoot, 1977). From these reports there appears to be a genuine species difference between stumptail and rhesus males within the laboratory testing environment.

The failure in previous studies to find significant hormonal regulation of coital activity in the stumptail has led the investigators to two, but not necessarily mutually exclusive, explanations. Slob et al. (1978a) suggested that the constant pairing and separation of animals may stimulate copulation, and that coital behavior serves as a "bonding function". Goldfoot (1977), in a general discussion of primate sexual behavior, has suggested that some instances of coital activity (especially when it occurs outside the time of ovulation) may be "social" in nature. These are interesting hypotheses and will be elaborated upon in the following discussion.

The results of this study of stumptails living in a social group demonstrated that the male had to be clearly dominant over the female in order for any aspect of sexual behavior to occur between them. It is suggested here that in the laboratory testing environment stumptail males and females are establishing their dominant and subordinate relationships through copulatory behavior. In support of this argument some additional information from the UCR colony is considered in conjunction with pertinent information from the laboratory studies.

When animals in this study were merged in 1972 to form a single social group, the males copulated with the "new" females regardless of their reproductive condition (Rhine, unpublished). Subsequent to this study, three females (each in a different phase of the menstrual cycle) were removed separately from this study group and placed in another, but unfamiliar group at UCR. Each time, the male in the new group achieved three to four ejaculations with the unfamiliar female in one-half hour of observation (Harvey, records 1977-1980). Similar high rates of ejaculation have been described by Goldfoot et al. (1978) when a stumptail female is first introduced to a male.

A recent report by Goldfoot et al. (1980) has described sexual behavior between stumptail females identical to the sexual repertoire of the male when tested in small groups. An analogous circumstance was observed with an all-female group at UCR. Females in the present study rarely exhibited even the most elementary sexual interest in each other. However, when three new stumptail females who had previously been housed individually at another facility, were placed together to form a social group, they displayed behaviors similar to sexual behaviors characteristic of the male. Within a few weeks the sexual behaviors diminished as the females became socially familiar with each other and their ranking system had been established (Harvey, records 1977-1980).

Apparently it makes little difference whether stumptails are housed individually or in a social group: when male-female and female-female stumptails meet for the first time, high levels of sexual behavior will ensue.

In contrast, brief separation (3 to 10 days) of females in this study group did not stimulate copulation upon re-entry, although threats, attacks, partial mounts and genital inspection were observed. From the nature and direction of the interaction it was clear that each animal was concerned with maintaining his/her previously established dominance position in the colony (Harvey, records 1977-1980).

To further illustrate the importance of the dominance relationship between the sexes it should be mentioned that during this study the four highest ranking males were simultaneously removed for six hours from the colony for dental treatment, leaving the three remaining lower ranking males with an opportunity to display sexual behavior without the interference of the more dominant males. Although the lower ranking males became sexually active, their preliminary sexual behavior and, ultimately, copulation were entirely directed to females lower ranking than themselves and not to the females who were dominant over them.

From the above discussion it appears that when stumptails are socially unfamiliar with each other and a rank-system is not well understood, then the display of coital behavior overrides the endocrine status of the female. However, when stumptails live in a social group and the dominant and subordinate relationships are recognized by all individuals, then the social and sexual behaviors are indeed responsive to the cycle phase of the female.

In a testing environment, without the presence of other males, the stumptail male has a unique oportunity to express his dominance through copulation. By establishing dominance, he gains potential reproductive access to that female. Females placed in temporary triads (Slob et al., 1978b) may also acquire as high a social rank as is possible in order to increase their reproductive success. Under such conditions, when the male is introduced, the high ranking female in order to maintain her rank might solicit the male while threatening the other females. This behavior on the part of the high ranking female probably stimulates the male to copulate with her, for by being dominant over her he is automatically dominant over the other females.

Females may have an interest in being high ranking, since there is evidence that offspring acquire rank proximate to the mother's rank. This has been documented for the Japanese macaque (Kawamura, 1958; Koyama, 1967), rhesus macaque (Sade, 1968; Missakian, 1972) and for stumptail macaques in this study. There is possibly another advantage, as Drickamer's (1974) study on rhesus macaques has suggested: higher ranking females have a definite reproductive advantage over the lower ranking females by giving birth more frequently, having better infant survival. Ultimately, their female offspring reproduce at an earlier age. It is possible that this is a general phenomenon for macaques. Consequently, males who have access to high ranking females have a definite reproductive advantage over males who are subordinate to such females.

Therefore, the coital behavior of stumptails in a testing environment is probably more "social" than sexual, as Goldfoot (1977) has suggested and the "bonds" (Slob et al., 1978a) which are being formed are in fact the dominant and subordinate relationships which are required for reproduction in socially living primates.

Available literature concerning macaques reveals that they live in complex social groups, with related females constituting the permanent membership of the group and males exchanging between groups for reproductive purposes (Drickamer and Vessey, 1972; Norikoshi and Koyama, 1975). Sexual behavior is not haphazard in Cercopithecinae, and while copulation can occur throughout the menstrual cycle, it is more prevalent around the time of ovulation (Goldfoot, 1977). In such systems there are rules of kinship, social rank, age and individual preferences (Goldfoot, 1977; Lancaster, 1979), which ultimately govern partner choice and timing of copulation.

This study on group-living stumptail macaques demonstrates how social factors interact with the female's reproductive condition to influence the display of social and sexual behaviors. The results of this study along with available information from the laboratory testing environment should be taken into consideration in future investigations concerning hormonal regulation of sexual behavior in Macaca arctoides, to provide a better understanding of the evolution of sexual behavior and reproduction in primates, including man.

ACKNOWLEDGEMENTS

This research was supported in part by the NIMH training program grant MH13006-01A1 and the Harry

Frank Guggenheim Foundation. Special acknowledgement goes to Ramon Rhine for making available the stumptail colony at the University of California, Riverside.

REFERENCES

Altmann, J. Observational study of behavior: Sampling methods. Behaviour 49, 227-267 (1974).

Baum, M.J., Slob, A.K., deJohn, F.H., and Westbroek, D.L. Persistence of sexual behavior in ovariectomized stumptail macaques following dexamethasone treatment or adrenalectomy. Hormones and Behavior 11, 323-347 (1978).

Brügenmann, S. and Grauwiler, J. Breeding results from an experimental colony of Macaca arctoides, in Medical Primatology, Part I. Karger, Basel (1972), pp. 216-226.

Bullock, D.W., Paris, C.A., and Goy, R.W. Sexual behavior, swelling of the sex skin and plasma progesterone in the pigtail macaque. J. Reprod. Fertil. 31, 225-236 (1972).

Czaja, J.A. and Bielert, C. Female rhesus sexual behavior and distance to a male partner: Relation to stage of the menstrual cycle. Arch. Sex. Behav. 4, 583-597 (1975).

Dixson, A.F., Everitt, J.B., Herbert, J., Rugenam, S.M., and Scruton, D.M. Hormonal and other determinants of sexual attractiveness and receptivity in rhesus and talapoin monkeys, in Symposium of the IVth International Congress of Primatology, Vol. 2: Primate Reproductive Behavior, C.H. Phoenix, ed. Karger, Basel (1973), pp. 36-63.

Drickamer, L.C. A ten year study of reproductive data for free-ranging Macaca mulatta. Folia Primatol. 21, 61-80 (1974).

Drickamer, L.C. and Vessey, S.H. Group changing in free-ranging rhesus monkeys. Primates 14, 359-368 (1973).

Eaton, G.G. and Resko, J.A. Ovarian hormones and sexual behavior in Macaca nemestrina. J. Comp. and Physiol. Psychol. 86, 919-925 (1974).

Everitt, B.J., Herbert, J., and Hamer, J.D. Sexual receptivity of bilaterally adrenalectomized female rhesus monkeys. Physiol. and Behav. 8, 409-415 (1972).

Goldfoot, D.A. Hormonal and social determinants of sexual behaviour in the pigtail monkey (Macaca nemestrina), in Normal and Abnormal Development of Brain and Behaviour, G.B.A. Stoelinga and J.J. van der Werff ten Bosch, eds. Leiden University Press, Leiden, Holland (1971), pp. 325-341.

Goldfoot, D.A. Sociosexual behaviors of nonhuman primates during development and maturity: Social and hormonal relationships, in Behavioral Primatology, Advances in Research and Theory, Vol. 1, A.M. Schrier, ed. Lawrence Erlbaum Assoc., Hillside, N.J. (1977), pp. 139-184.

Goldfoot, D.A., Wiegand, S.J., and Scheffler, G. Continued copulation in ovariectomized adrenal-suppressed stumptail macaques (Macaca arctoides). Hormones and Behavior 11, 88-99 (1978).

Goldfoot, D.A., Westerborg-van-loon, H., Groeneveld, W., and Slob, A.K. Behavioral and physiological evidence of sexual climax in the female stump-tailed macaques (Macaca arctoides). Science 208, 1477-1478 (1980).

Harvey, N.C., Rhine, R.J. and Bunyak, S.C. Weights and heights of stump-tailed macaques (Macaca arctoides) living in colony groups. J. Med. Primatol. 8, 372-376 (1979).

Johnson, D.F. and Phoenix, C.H. The hormonal control of female sexual attractiveness, proceptivity, and receptivity in rhesus monkeys. J. Comp. Physiol. Psychol. 90, 473-483 (1976).

Kawamura, S. Matriarchal social ranks in the Minoo-B troop: A study of the rank system of Japanese monkeys. Primates 1, 148-156 (1958).

Koyama, N. On dominance rank and kinship of a wild Japanese monkey troop in Arashyama. Primates 8, 189-216 (1967).

Lancaster, J. Sex and gender in evolutionary perspective, in Human Sexuality: A Comparative and Developmental Perspective, H.A. Katchadourian, ed. University of California Press, Berkeley and Los Angeles, California (1979), pp. 51-80.

Macdonald, G.J. Reproductive patterns of three species of macaques. Fertil. Steril. 22, 373-377 (1971).

Michael, R.P. and Bonsall, R.W. Periovulatory synchronization of behaviour in male and female rhesus monkeys. Nature (Lond.) 265, 463-465 (1977).

Michael, R.P. and Zumpe, D. Sex initiating behaviour by female rhesus monkeys (Macaca mulatta) under laboratory conditions. Behaviour 36, 168-186 (1970).

Michael, R.P., Zumpe, D., Keverne, E.B., and Bonsall, R.W. Neuroendocrine factors in the control of primate behaviour. Recent Progress in Hormone Research 28, 665-706 (1972).

Missakian, E.A. Genealogical and cross-genealogical dominance relationships in a group of free-ranging rhesus monkeys (Macaca mulatta) on Cayo Santiago. Primates 13, 169-180 (1972).

Norikoshi, K. and Koyama, N. Group shifting and social organization among Japanese monkeys, in Proceedings from the Symposia of the 5th Congress of the International Primatological Society, S. Kondo, M. Kawai, A. Ehara, and S. Kawamura, eds. Science Press, Tokyo (1975).

Perachio, A.A., Alexander, M., and Marr, L.D. Hormonal and social factors affecting evoked sexual behavior in rhesus monkeys. Am. J. Phys. Anthrop. 38, 227-232 (1973).

Rhine, R.J. Changes in the social structure of two groups of stumptail macaques (Macaca arctoides). Primates 13, 181-194 (1972).

Rhine, R.J. and Kronenwetter, C. Interaction patterns of two newly formed groups of stumptail macaques (Macaca arctoides). Primates 13, 19-33 (1972).

Sade, D.S. Inhibition of son-mother mating among free-ranging rhesus monkeys. Science and Psychoanalysis 12, 19-38 (1968).

Slob, A.K., Wiegand, S.J., Goy, R.W., and Robinson, J.A. Heterosexual interactions in laboratory-housed stumptail macaques (Macaca arctoides): Observations during the menstrual cycle and after ovariectomy. Hormones and Behavior 10, 191-211 (1978a).

Slob, A.K., Baum, M.J., and Schenck, P.E. Effects of the menstrual cycle, social grouping and exogenous progesterone on heterosexual interaction in laboratory housed stumptail macaques (Macaca arctoides). Physiol. and Behav. 21, 915-921 (1978b).

Slob, A.K., Ooms, M.P., and Vreeburg, J.T.M. Annual changes in serum testosterone in laboratory housed male stumptail macaques (Macaca arctoides). Biology of Reproduction 20, 981-984 (1979).

Trollope, J. and Blurton Jones, N.G. Aspects of reproductive behavior in Macaca arctoides. Primates 16, 191-205 (1975).

Wallen, K. and Goy, R.W. Effects of estradiol benzoate, estrone, and propionates of testosterone or dihydrotesterone and sexual and related behaviors of ovariectomized rhesus monkeys. Hormones and Behavior 9, 228-248 (1977).

Wilks, J.W. Endocrine characterization of the menstrual cycle of the stumptailed monkey (Macaca arctoides). Biol. of Reprod. 16, 479-485 (1977).

Chapter 7
Effects of Cyproterone Acetate on Social and Sexual Behavior in Adult Male Laboratory Housed Stumptailed Macaques *(Macaca Arctoides)*

A.K. Slob, P.E. Schenck, and H. Nieuwenhuijsen

INTRODUCTION

Cyproterone acetate (CA; Androcur^R) is a synthetic drug with antiandrogenic and antigonadotrophic properties (Neumann et al., 1977). Important indications for the clinical use of this drug are advanced prostatic carcinoma (e.g., Bartsch et al., 1977; Jacobi et al., 1979) and hirsutism and/or acne in women (Hammerstein et al., 1975; Abrahamsson et al., 1979; Underhill and Dewhurst, 1979; Hammerstein et al., 1980). Cyproterone acetate is also used in psychiatry for the treatment of "hypersexuality and sexual deviations" in men (e.g., Horn, 1974). In this context two quotations of German investigators are of interest: "..., cyproterone acetate has been used clinically since 1967 to decrease libido and potency in subjects with sexual instincts that are either excessive or abnormal in mode." (Horn, 1977, p. 352); and "The therapy of pathological hypersexuality is one of the established indications for cyproterone acetate." (Neumann et al., 1980, p. 151).

Despite the increasing 'popularity' of CA in several European countries, well designed double-blind clinical studies on the effects of CA on human sexual behavior are lacking. Warnings against prescription of CA in this behavioral contest are rare (Werff ten Bosch, 1973).

Many studies on small laboratory species such as the rat and guinea-pig have failed to reveal diminishing effects of CA on sexual behavior (Zucker, 1966; Whalen and Edwards, 1968; see also review by Neumann and Steinbeck, 1974). One study reported stimulatory effects of CA on sexual behavior of rats (Bloch and Davidson, 1971). To our knowledge, only three studies have reported some, but not a very striking reduction in male copulatory behavior: in the Japanese quail (Adkins and Mason, 1974), in the rabbit (Ågmo, 1975), and in the dog (Schmidtke and Schmidtke, 1968). The only preliminary study on the effects of CA on sexual behavior in a higher nonhuman primate (adult male rhesus monkey, Plant and Michael, 1973; Michael et al., 1973) revealed equivocal results. One male rhesus stopped copulating during CA administration, one male's ejaculatory performance seemed virtually unaffected, and a third male exhibited marked difficulty in obtaining intromission during treatment.

Although there seems to be considerable current clinical interest in the use of CA, there are no systematic studies on the effects of this drug in higher non-human primates. Therefore, we have carried out two double-blind experiments and one placebo-controlled study on the effects of CA on heterosexual and social behavior in adult male stumptailed macaques.

The stumptailed macaque seemed to be a good experimental animal for the purpose of this study. This macaque species displays high levels of sexual behavior in (short) heterosexual laboratory pair tests (Goldfoot et al., 1975; Slob and Nieuwenhuijsen, 1980). Moreover, this behavior seems independent of endogenous sex hormones in the female partner (Slob et al., 1978a; Slob et al., 1978b; Baum et al., 1978).

GENERAL METHOD

Animals

Ten adult male and ten adult female stumptailed macaques (M. arctoides) were used in these experiments. In experiments I and II the female partners received silastic implants containing crystalline estradiol (for details see Baum et al.,

1978) in order to prevent pregnancy. In experiment III the female partners were tubal-ligated two months before the start of behavior tests. All animals were over nine years of age and with the exception of male 444, had been born in the wild. They were housed in individal cages in a colony room where they had audio-visual contact with other males and females at the Primate Center TNO, Rijswijk, The Netherlands. The colony room was isolated from the room where the animals were tested. Lights in the colony were out between 19.00 and 08.00 hr. Animals were fed rice, monkey chow, and fruit in the morning prior to behavioral testing as well as late each afternoon.

Behavior Tests

In each experiment heterosexual pairs of monkeys were selected and subsequently tested in a balanced sequence four mornings per week (Monday through Thursday). Males were tested twice a week with 24 to 48 hr between tests. In experiments I and II the female partners rotated in such a way that each male was paired with a different female partner for each test. Experiment III was conducted with fixed pairs.

Tests were carried out in a large cage (1.65 x 1.65 x 2.5 m) which had a Plexiglass front. The animals' interaction was scored by two or three observers from behind a one-way vision screen using a 20-channel Esterline Angus event recorder. The interaction of a pair was recorded for 10 minutes (Experiments I and II) or 30 minutes (Experiment III), using the behavioral items previously described for stumptails (Goldfoot et al., 1975; Baum et al., 1978).

Hormone Assays

Samples of five ml blood were obtained on Fridays (at least 48 hr after a previous injection) from the femoral vein or artery while the males were lightly anesthetized with 30 mg ketamine hydrochloride each. The blood was allowed to clot and the serum subsequently stored at $-20^{\circ}C$ prior to the hormone determination.

Testosterone was estimated using the radioimmunoassay, with aluminum oxide chromatography, first described by Verjans and colleagues (1973). Serum cyproterone acetate content was estimated by B. Nieuweboer (Research Laboratories of Schering AG, Berlin/Berghamen) using a specific and sensitive radioimmunoassay first described by Nieuweboer and Lubke (1977). Serum cortisol was estimated by Dr. F.H. de Jong (Department of Biochemistry, Erasmus University, Rotterdam) using a radioimmunoassay first described by de Jong and van der Molen (1972).

Physical Measures

Only in experiment III were physical measures taken regularly. While anesthetized males were weighed the length of each testis was measured with calipers. Testicular size was subsequently estimated by palpating each testis and comparing the impression with a set of 23 plastic testis models that ranged in size from 1 to 40 ml. All testis measures were taken by two investigators independently; mean values were used for analysis.

Cyproterone Acetate Treatment

CA was injected intramuscularly once, twice, or three times per week. A mixture of benzyl benzoate and castor oil was used as a solvent, so that 1 ml of the solution contained 100 mg CA, 619 mg benzyl benzoate, and 353 mg castor oil. Sterile ampoules with CA or with the vehicle were prepared under the supervision of Professor B. Lenstra, pharmacist at the Academic Hospital Dykzigt, Rotterdam. For the double-blind experiments (I and II) the ampoules were coded A and B. The investigators were informed after the experiments had been completed.

EXPERIMENTS I AND II

Methods

Initially two double-blind experiments were carried out. Experiment I: 'normal' dose CA, between March and July 1978 (50 mg CA i.m./wk/male; eight males: 136, 318, 360, 374, 392, 429, 444, 465). Experiment II: 'higher' dose CA, between October 1978 and February 1979 (100 mg CA i.m./wk/male; six males: 136, 360, 392, 429, 444, 465). Beginning three weeks after the onset of the injections (CA or vehicle) behavior tests took place for three weeks (six tests/male). Injections and tests were then stopped for four weeks. This was followed by another six weeks of injections, during the last three of which behavior tests were carried out again. During the last six weeks the injection groups were reversed (criss-cross design).

Results and Discussion

Behavioral Data

Analysis of variance (ANOVA, two factorial block design) of the behavioral data revealed only one significant effect of CA, an increase in male grooming time in Experiment II ($F(1/55= 4.53$, $p < 0.05$) (Table 1). No other male or female social or sexual behavior was affected by CA administration to the male.

Overall, there was less sexual activity, but more male and female grooming in Experiment II than in Experiment I. We have no explanation for these findings other than the possibility that 'season' may have played a role.

Physical Data

Body weights were not affected by CA treatment (Table 2). The serum CA levels clearly indicate that significant amounts of CA were circulating in the animals. The finding that serum CA levels

MALE/FEMALE BEHAVIORS	EXPERIMENT I (March - July 1978)		EXPERIMENT II (October 1978-February 1979)	
	Cyproterone acetate (50 mg/wk/♂)	Oil vehicle (0.5 ml/wk/♂)	Cyproterone acetate (100 mg/wk/♂)	Oil vehicle (2 x 0.5 ml/wk/♂)
♂ Approach	5.2 ± 1.4	4.8 ± 0.6	5.4 ± 1.9	4.8 ± 1.4
Sex contact	4.7 ± 0.4	4.7 ± 0.4	4.8 ± 1.8	3.2 ± 0.5
Investigation ♀ genitalia	5.9 ± 0.7	6.0 ± 0.6	4.6 ± 1.4	4.3 ± 0.8
Mount	4.4 ± 0.5	4.8 ± 0.5	3.7 ± 1.8	2.4 ± 0.8
Intromission	2.0 ± 0.3	2.0 ± 0.3	1.4 ± 0.6	1.0 ± 0.4
Ejaculation	1.4 ± 0.2	1.2 ± 0.6	0.9 ± 0.3	0.7 ± 0.3
*Thrusts to 1st ejaculation	38.9 ± 7.1	59.2 ± 17.5	50.5 ± 23.4	78.8 ± 32.2
Yawn	1.0 ± 0.6	1.2 ± 0.6	0.6 ± 0.3	0.8 ± 0.5
Groom (min)	1.1 ± 0.4	0.7 ± 0.3	1.9 ± 1.0^Φ	0.8 ± 0.4
Threat and Aggression	0.3 ± 0.1	0.4 ± 0.2	0.7 ± 0.6	1.0 ± 0.8
*Latency to 1st mount (min)	0.3 ± 0.1	0.3 ± 0.1	0.8 ± 0.5	0.5 ± 0.3
*Latency to 1st intro (min)	1.1 ± 0.3	1.6 ± 0.7	0.6 ± 0.2	0.4 ± 0.1
*Latency to 1st ejac (min)	1.7 ± 0.5	1.8 ± 0.3	1.6 ± 0.4	1.7 ± 0.5
*Post 1st ejac intro (min)	1.0 ± 0.2	1.0 ± 0.1	1.4 ± 0.6	0.9 ± 0.2
*Interval 1st ejac - next mt (min)	4.0 ± 0.4	5.0 ± 0.6	5.8 ± 1.7	6.3 ± 0.4
Sit next (min)	0.1 ± 0.03	0.2 ± 0.1	0.2 ± 0.1	0.2 ± 0.1
Number tests c̄ ejaculation (max. = 6)	4.5 ± 0.7	4.5 ± 0.7	2.8 ± 1.0	2.3 ± 1.0
♀ Approach	1.5 ± 0.3	1.6 ± 0.2	1.9 ± 0.4	2.2 ± 0.5
Presentation	7.6 ± 1.0	7.5 ± 0.5	9.0 ± 2.0	7.7 ± 0.6
Refuse to contact,mt,intro	1.4 ± 0.4	1.7 ± 0.6	1.2 ± 0.4	1.3 ± 0.3
Groom (min)	1.7 ± 0.4	2.1 ± 0.3	2.5 ± 0.6	3.2 ± 0.4

*For responders only
^Φdifferent from oil vehicle ($p < 0.05$)

TABLE 1. Various male and female stumptailed monkey behaviors (mean ± SEM/test) during two double-blind cyproterone acetate experiments. Experiment I: 'normal' dose of CA (50 mg i.m./wk/♂), total 8 males. Experiment II: 'higher' dose of CA (2 x 50 mg i.m./wk/♂), total 6 males. Behavior tests consisted of 10 min heterosexual pair tests.

in Experiment II were not significantly higher than in Experiment I, although the animals received two injections per week in Experiment II, needs some comment. During the second phase of Experiment II we noticed crystals in ampoules coded A, which indicated that the CA was perhaps no longer completely dissolved. There was nothing that could be done at that time other than to finish the experiment. Afterward the serum CA levels of the males in Experiment II turned out to be comparable with the levels of the males in Experiment I and the data are therefore still presented.

ANOVA (2 factorial block design) of the testosterone data revealed that during CA treatment serum testosterone levels remained constant (Experiment I: X = 7.1 ng/ml; Experiment II: X = 3.8 ng/ml). In the vehicle condition serum testosterone levels were significantly higher than during CA treatment. (Experiment I: X = 12.6 ng/ml; Experiment II: X = 7.9 ng/ml). Furthermore, in Experiment I there was a significant increase in serum testosterone during behavior tests only in the control condition ($F(2/35)=4.92$, p-0.05). In Experiment II there was the same trend, although not statistically significant. Comparisons of testosterone data between identical experimental conditions showed lower levels of serum testosterone in Experiment II. There is no easy explanation for this finding. An explanation could be sought in the period of the year ("season") in which the two experiments were conducted. We have earlier reported that laboratory housed male stumptailed monkeys showed higher serum testosterone levels during spring and summer than during fall and winter (Slob et al., 1979).

EXPERIMENT III

The results of the previous experiments suggested that in the laboratory housed stumptailed male monkey six weeks of CA treatment had no effects on sexual behavior. Even so, some criticism was possible. For example, the CA doses might have been too small and/or given over too short a period of

182 SLOB ET AL.

Table 2. Mean (± SEM) serum levels of testosterone and cyproterone acetate (ng/ml) and body weights (kg) of male stumptailed macaques during two double-blind CA-experiments. Experiment I: 'normal' dose of CA (50 mg i.m./wk/♂), total 8 males. Experiment II: 'higher' dose of CA (2 x 50 mg i.m./wk/♂), total 6 males. During weeks with behavior tests the males were subjected twice to a 10 min heterosexual pair test.

TREATMENT	HORMONES/ BODY WEIGHT	BEFORE START EXPERIMENTS	WEEKS FOLLOWING START OF INJECTIONS					2 WKS AFTER LAST INJ. & BEH.TEST
			1 NO BEHAVIOR TESTS	2 NO BEHAVIOR TESTS	3 1 WK BEH.T.	4 2 WKS BEH.T.	5 3 WKS BEH.T.	
EXPERIMENT I 50 mg/wk/♂ (March – July 1978)	Testosterone	8.0 ± 1.3			6.9 ± 1.2	7.7 ± 1.2	6.7 ± 1.4	6.8 ± 1.0
	Cyproterone acetate	N.D.			47.0 ± 4.7	63.5 ± 7.0	66.6 ± 11.7	N.D.
	Body weight	10.2 ± 0.4			9.8 ± 0.5	9.7 ± 0.5	9.5 ± 0.5	9.3 ± 0.4
Oil vehicle .5 ml/wk/♂	Testosterone	7.5 ± 1.3			11.1 ± 1.4	11.7 ± 1.0	14.4 ± 2.7	13.1 ± 2.0
	Cyproterone acetate	N.D.			N.D.	N.D.	N.D.	N.D.
	Body weight	9.8 ± 0.4			10.0 ± 0.4	9.8 ± 0.4	9.7 ± 0.4	9.6 ± 0.4
EXPERIMENT II 100 mg/wk/♂ (October 1978 – February 1979)	Testosterone		4.5 ± 0.1	3.7 ± 0.6	3.3 ± 0.4	4.8 ± 1.2	2.8 ± 0.3	
	Cyproterone acetate		63.8 ± 18.6	68.3 ± 25.9	74.0 ± 19.5	58.7 ± 14.7	52.0 ± 10.0	
	Body weight		9.0 ± 0.8	9.1 ± 0.7	9.2 ± 0.7	9.0 ± 0.7	8.9 ± 0.7	
Oil vehicle 2x .5 ml/wk/♂	Testosterone		6.7 ± 1.3	6.8 ± 0.9	10.9 ± 2.3	7.4 ± 1.7	7.5 ± 1.4	
	Cyproterone acetate		N.D.	N.D.	N.D.	N.D.	N.D.	
	Body weight		8.9 ± 0.5	8.9 ± 0.4	9.2 ± 0.4	9.1 ± 0.4	9.1 ± 0.4	

time; tests of ten minutes might have been too short; rotation of female partners might have been too 'stimulating' for the males. Because of these considerations, a third experiment was performed.

Method

Seven heterosexual pairs of stumptailed monkeys (seven males and seven females), which copulated readily in pretests, were used. The experiment consisted of two episodes, one between September and December 1979, the other between April and July 1980. Each lasted 14 weeks. Behavior tests (30 minutes per test) were begun two weeks prior to the start of the injections and were continued (2x/wk/pair) throughout the period of injections. Injections (i.m.) were given for 12 weeks. Cyproterone acetate was administered in increasing doses; six weeks 100 mg/wk, two weeks 150 mg/wk, and finally four weeks 210 mg/wk. This was not a double-blind experiment.

During the first episode of 14 weeks four males (135, 360, 429, 465) received CA and the remaining three males (392, 444, 453) vehicle injections. During the second episode the injection groups were reversed, so that four males received vehicle and three males CA.

Results and Discussion

Physical Data

Body weights were not affected by CA treatment (Table 3). Serum CA levels were considerably higher than in the earlier experiments, and within the present experiment they were dose-dependent (one-way ANOVA, block design, $(F(2/12)=13.94, p<0.01)$. Subsequent analysis with the simple main effects method showed that during CA treatment serum testosterone levels were significantly lower than they were before the tests and/or injections were started ($p<0.01$). There were no changes in serum testosterone levels during vehicle treatment ($F(5/66)= 0.45$, ns). Serum cortisol levels were

TREATMENT	PHYSICAL AND HORMONAL DATA	BEFORE START EXPERIMENT	2 WEEKS AFTER START BEHAVIOR TESTS NO INJECTIONS	WEEKS FOLLOWING START OF INJECTIONS AND BEHAVIOR TESTS				3+ OR 7 WEEKS AFTER LAST INJECTION AND BEHAVIOR TEST
				6	8		12	
CYPROTERONE ACETATE	Body weight	8.8 ± 0.8	8.8 ± 0.8	8.4 ± 0.8	8.6 ± 0.6		8.4 ± 0.6	8.5 ± 0.5
	Testislength		48 ± 2		42 ± 2		40 ± 2	46 ± 2
	Testisvolume		33 ± 5	26 ± 5	25 ± 7		22 ± 6	32 ± 6
	Testosterone	12.3 ± 2.6	12.2 ± 2.9	4.2 ± 0.6	3.8 ± 0.9		3.7 ± 1.0	5.2 ± 0.9+
	Cyproterone acetate	N.D.	N.D.	120 ± 15	193 ± 19		297 ± 34	N.D.+
	Cortisol	19.2 ± 2.4	17.9 ± 2.5	19.6 ± 2.7	19.0 ± 3.7		16.2 ± 1.4	—
OIL VEHICLE	Body weight	8.9 ± 0.6	9.0 ± 0.5	9.1 ± 0.5	9.1 ± 0.5		8.6 ± 0.5	8.5 ± 0.5
	Testislength		48 ± 2		48 ± 2		47 ± 2	48 ± 2
	Testisvolume		31 ± 6	30 ± 7	29 ± 7		29 ± 7	37 ± 1
	Testosterone	7.4 ± 0.9	8.9 ± 2.4	6.8 ± 1.0	8.9 ± 1.9		7.2 ± 1.5	6.9 ± 1.2+
	Cyproterone acetate	N.D.	N.D.	N.D.	N.D.		N.D.	N.D.+
	Cortisol	19.2 ± 2.5	17.0 ± 2.5	17.6 ± 2.3	16.7 ± 3.3		24.2 ± 3.0	—

Table 3. Various physical data (body weight -kg-, testislength -mm-, testisvolume -ml-) and serum endocrine data (testosterone -ng/ml-, cyproterone acetate -ng/ml-, cortisol -ng/100ml-) of 7 male stumptailed macaques during CA injections (no: 6 weeks 100 mg/wk/♂; 2 weeks 150 mg/wk/♂; 4 weeks 210 mg/wk/♂) and during oil vehicle injections (Experiment III: cross-cross experimental design). Values are mean ± SEM.

not affected by CA treatment ($F(1/54)=0.14$, ns). This latter finding agrees with results reported by Smals and colleagues (1978) that cyproterone acetate had no effect on the pituitary-adrenal system of hirsute women.

During CA administration the testes became smaller, both in length and in volume. For the analysis one animal (465) with extremely small testes was excluded. A significant treatment X duration interaction (ANOVA, 2 factorial block design, $F(3/35)=7.17$, $p<0.01$) indicated that after eight and 12 weeks of CA injections testicular length was decreased compared to pre-injection and post-injection values. Virtually the same results were obtained for estimated testes volume.

Behavioral Data

Initially, ANOVA revealed that for various behaviors there were already significant differences between the treatment groups during the pre-treatment tests. Therefore, difference measures (cf. Kirk, 1969, p. 487) were used, with the pre-treatment data as the initial scores. Thus an ANOVA (2 factorial design) on difference measures was done. In no instance did ANOVA reveal a significant effect of duration, nor was there a significant interaction. The presented p-values in Tables 4 and 5 indicate a significant treatment effect.

Male sexual behavior was significantly affected by CA treatment. There was an increase in frequency of sex exploration (investigation of female genitalia), mounting, intromission and ejaculation. The number of pelvic thrusts to first ejaculation, as well as frequency of sex contact, was relatively constant during CA treatment, whereas there was a decrease during vehicle injections. The post-first-ejaculation-intromission time was significantly shorter during CA treatment. Latencies to first mount and to first ejaculation were not affected by cyproterone acetate.

Some other male social behaviors were also affected by CA administration. There was a striking

BEHAVIOR	TREAT-MENT	no injections 2	WEEKS FOLLOWING BEGINNING OF BEHAVIOR TESTS							ANOVA difference measures (CA vs OIL)
			100 mg CA or oil/wk/♂ 4	6	8	150 mg CA or oil/wk/♂ 10	210 mg CA or oil/wk/♂ 12	14		
Approach	CA	7.2 + 2.8	6.8 + 3.0	6.1 + 2.6	6.1 + 3.0	5.1 + 2.3	7.1 + 4.3	6.5 + 3.5	$p < 0.01$
	oil	9.3 + 2.9	5.8 + 2.0	5.5 + 1.5	4.1 + 0.9	4.1 + 1.1	3.3 + 1.0	2.4 + 1.1	
Sex contact	CA	4.8 + 0.9	4.7 + 1.4	4.4 + 1.1	4.3 + 0.9	4.2 + 0.8	4.1 + 1.0	4.4 + 0.9	$p < 0.01$
	oil	5.6 + 1.0	4.7 + 1.6	4.5 + 1.2	3.6 + 0.9	3.2 + 1.1	3.2 + 0.8	2.8 + 1.0	
Invest. ♀ genitalia	CA	1.5 + 0.4	1.8 + 0.6	1.8 + 0.7	2.2 + 0.6	2.5 + 0.8	2.4 + 0.8	1.9 + 0.7	$p < 0.01$
	oil	1.4 + 0.3	2.2 + 0.8	1.4 + 0.3	1.1 + 0.3	1.4 + 0.4	1.1 + 0.3	1.4 + 0.6	
Mount	CA	3.7 + 0.8	4.2 + 1.2	4.2 + 1.1	4.2 + 0.9	4.4 + 0.8	4.0 + 0.9	4.2 + 0.7	$p < 0.01$
	oil	5.0 + 0.7	3.5 + 1.2	3.8 + 1.0	3.5 + 0.7	2.6 + 0.8	3.1 + 0.7	2.8 + 1.0	
Intromission	CA	2.7 + 0.6	2.9 + 0.7	3.0 + 0.6	3.2 + 0.8	3.6 + 0.5	3.5 + 0.7	3.6 + 0.6	
	oil	3.7 + 0.9	2.6 + 0.9	3.0 + 0.9	2.6 + 0.8	2.2 + 0.8	2.8 + 0.7	2.2 + 0.7	
Ejaculation	CA	2.0 + 0.4	2.6 + 0.6	2.8 + 0.7	2.6 + 0.7	2.8 + 0.7	2.9 + 0.7	2.9 + 0.6	$p < 0.01$
	oil	2.6 + 0.8	2.1 + 0.8	2.5 + 0.8	1.9 + 0.6	1.8 + 0.7	2.2 + 0.6	1.6 + 0.5	
+Thrusts to 1st ejacul.∅	CA	57.1 + 20.1	58.0 + 13.0	47.3 + 11.1	58.8 + 22.4	58.0 + 17.7	45.2 + 11.1	54.8 + 16.5	$p < 0.01$
	oil	51.0 + 15.0	36.9 + 6.6	49.2 + 15.0	39.7 + 6.7	39.8 + 6.5	37.6 + 7.9	34.9 + 5.3	
+Latency to 1st mt (min)	CA	0.4 + 0.1	2.5 + 2.2	0.6 + 0.2	1.6 + 0.8	1.0 + 0.5	0.4 + 0.1	0.5 + 0.2	
	oil	1.2 + 0.4	2.0 + 1.4	1.3 + 0.4	0.8 + 0.2	1.6 + 0.6	1.5 + 0.7	3.1 + 1.7	
+Latency to 1st ej (min)∅	CA	1.6 + 0.5	1.5 + 0.4	1.5 + 0.5	4.0 + 2.0	3.4 + 1.3	1.1 + 0.2	2.4 + 0.8	
	oil	2.4 + 0.5	3.8 + 1.3	3.0 + 0.5	4.1 + 1.7	2.6 + 0.6	2.3 + 1.2	4.5 + 2.6	
+Post 1st ej. introm (min)∅	CA	0.9 + 0.3	0.9 + 0.4	0.5 + 0.1	0.6 + 0.2	0.7 + 0.2	0.6 + 0.2	0.7 + 0.2	$p < 0.01$
	oil	0.6 + 0.2	0.7 + 0.2	0.8 + 0.3	0.8 + 0.3	0.8 + 0.3	0.8 + 0.3	0.7 + 0.1	
Yawn	CA	1.6 + 1.0	2.1 + 1.3	0.6 + 0.4	0.6 + 0.3	0.4 + 0.2	0.2 + 0.1	0.3 + 0.1	$p < 0.01$
	oil	3.6 + 1.7	2.1 + 0.8	1.3 + 0.8	1.5 + 0.8	1.3 + 0.2	1.0 + 0.4	0.8 + 0.4	
Groom (min)	CA	4.8 + 1.8	7.5 + 2.2	9.7 + 2.7	9.8 + 2.6	11.3 + 3.2	10.2 + 3.1	10.6 + 3.0	$p < 0.01$
	oil	5.0 + 1.9	3.4 + 2.1	4.9 + 2.3	4.6 + 1.9	4.8 + 2.2	6.8 + 2.6	5.1 + 2.3	
Groom solic.	CA	0.8 + 0.3	0.5 + 0.3	0.3 + 0.1	0.4 + 0.3	0.3 + 0.2	0.1 + 0.0	0.2 + 0.1	
	oil	1.2 + 0.4	0.6 + 0.4	1.0 + 0.4	0.7 + 0.4	0.7 + 0.5	1.1 + 0.5	0.5 + 0.4	
Sit next(min)	CA	0.8 + 0.4	0.5 + 0.1	0.4 + 0.2	0.5 + 0.2	0.3 + 0.2	0.3 + 0.2	0.3 + 0.2	
	oil	1.3 + 0.4	1.6 + 1.1	1.1 + 0.5	0.9 + 0.4	1.0 + 0.6	1.3 + 0.5	1.3 + 0.9	

+ *for responders only*
∅ *without males 360 & 444*

Table 4. Various behaviors (mean + SEM) of 7 male stumptailed macaques tested during 3 different dosages of $\overline{CA}$ and during oil vehicle injections (Experiment III: criss-cross experimental design). Data presented per block of 2 weeks (=

BEHAVIOR	TREAT-MENT ♂ PART-NER	WEEKS FOLLOWING START OF BEHAVIOR TESTS WITH ♂ PARTNERS RECEIVING							ANOVA difference measures (CA vs OIL)
		no inject. 2	100 mg CA or oil/wk/♂ 4	100 mg CA or oil/wk/♂ 6	150mg CA or oil/wk/♂ 8	150mg CA or oil/w/♂ 10	210 mg CA or oil/wk/♂ 12	210 mg CA or oil/wk/♂ 14	
Approach	CA	1.3 + 0.2	1.3 + 0.4	1.5 + 0.6	1.6 + 0.4	1.0 + 0.6	1.6 + 0.9	1.2 + 0.4	
	oil	3.2 + 1.0	3.1 + 1.0	3.5 + 1.2	3.3 + 1.5	3.2 + 1.2	3.2 + 1.2	2.2 + 1.1	
Voluntary Present	CA	2.8 + 0.7	1.9 + 0.8	1.4 + 0.8	1.5 + 0.4	1.2 + 0.2	0.6 + 0.1	1.1 + 0.4	$p < 0.01$
	oil	2.6 + 0.8	2.4 + 0.8	3.0 + 1.4	2.4 + 0.7	3.3 + 1.4	3.3 + 1.4	2.0 + 0.8	
Male induced Present	CA	3.6 + 0.4	3.7 + 1.0	4.0 + 1.0	4.3 + 1.0	4.3 + 0.9	4.2 + 0.7	4.2 + 1.0	$p < 0.01$
	oil	6.4 + 1.3	4.8 + 1.4	4.1 + 0.8	3.1 + 0.7	3.3 + 1.1	2.8 + 0.7	2.5 + 1.0	
Groom (min)	CA	12.2 + 4.1	10.7 + 4.0	12.6 + 3.6	11.8 + 3.6	11.8 + 3.7	12.3 + 4.0	11.8 + 4.0	$p < 0.01$
	oil	10.5 + 3.2	13.5 + 3.9	12.3 + 4.0	11.8 + 3.9	12.8 + 4.2	11.6 + 3.9	13.2 + 4.5	

Table 5. Four behaviors (mean + SEM) of female partners of the 7 male stumptailed macaques during CA injections (3 different dosages) and during oil vehicle injections (Experiment III: criss-cross experimental design). Data presented per block of 2 weeks (= 4 behavior tests).

188 SLOB ET AL.

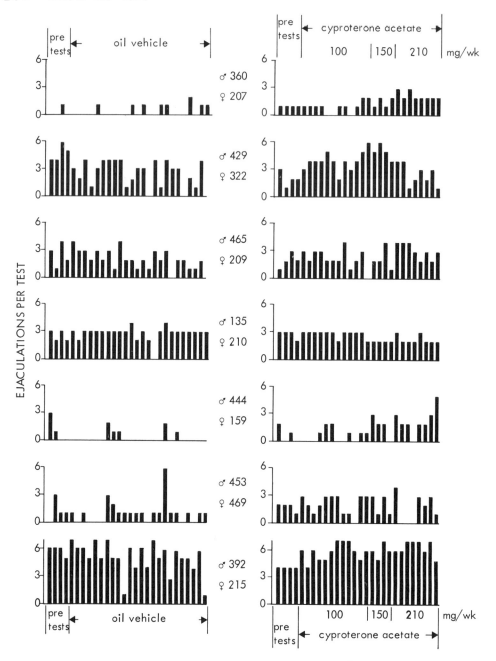

Figure 1. Number of ejaculations per test (30 min behavior test; two tests/wk) of 7 pairs of stumptailed macaques during 12 weeks with CA injections in three dosages and during 12 weeks with oil vehicle injections. Note the inter-pair differences and intra-pair consistencies.

CYPROTERONE ACETATE AND SOCIO-SEXUAL BEHAVIOR 189

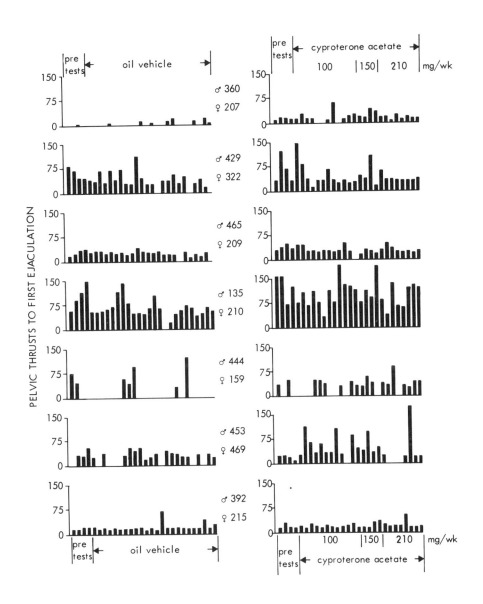

Figure 2. Number of pelvic thrusts to the first ejaculation per test exhibited by seven male stumptailed macaques (two 30 min behavior tests/pair/wk) during 12 weeks with CA injections in three dosages and during 12 weeks with oil vehicle injections. Note the inter-pair differences and intra-pair consistencies.

increase in groom duration, although frequency of groom solicit and sit next duration were not affected by CA. Frequency of yawning decreased in both conditions, but more so during oil vehicle injections. Frequencies of threat and aggression were extremely low and are therefore not presented.

Female behaviors that were affected by CA treatment of their male partners can be found in Table 5. The frequencies of voluntary presents (present near and present at the distance) were decreased, but induced presents (present to male approach and present to contact) were increased during CA. Groom duration was somewhat lower during CA. Frequencies of negative behaviors (e.g., refusals to contact, to being mounted, or to allow intromission) were extremely low and are not presented.

There were rather striking differences in socio-sexual behaviors between the seven pairs of animals. Furthermore, it appeared that for each pair the behavioral interaction was rather stereotyped. Therefore, individual data are presented in Figures 1 through 5.

If we consider ejaculation frequency (Figure 1) we notice first of all an increase in six out of seven pairs during CA treatment. On the other hand, the individual patterns are rather constant: pair 392-215 usually had between four and seven ejaculations per test, whereas pair 135-210 usually ranged between two and three ejaculations per test, and pair 444-159 ranged between zero and three. The same phenomena can be observed for the number of pelvic thrusts to first ejaculation (Figure 2): pair 392-215 usually had around 20-25 thrusts, but pair 135-210 was more variable with 60 to 90 thrusts. The other five pairs had intermediate values.

Similar inter-pair differences and intra-pair consistencies can be seen for latency to first ejaculation (Figure 3) and the time the pair maintained an intromission following the first ejaculation (Figure 4). Not only was sexual behavior 'pair-specific', but also social behavior such as grooming followed this trend (Figure 5). There we notice first of all an increase in male grooming in

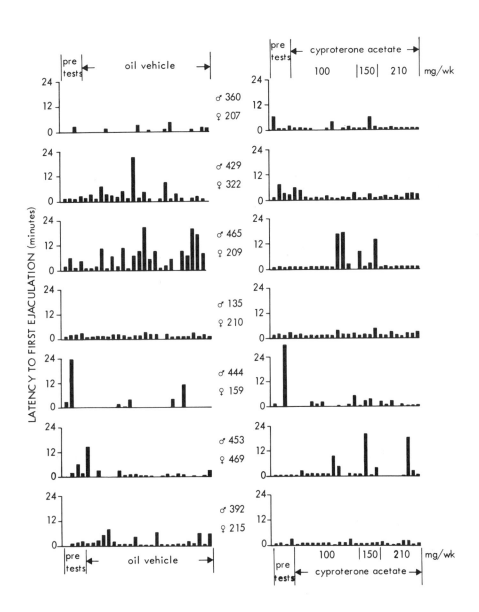

Figure 3. Latency (in min) to first ejaculation per test of seven pairs of stumptailed macaques (two 30 min behavior tests/pair/wk) during 12 weeks with CA injections in three dosages and during 12 weeks with oil vehicle injections. Note the inter-pair differences and intra-pair consistencies.

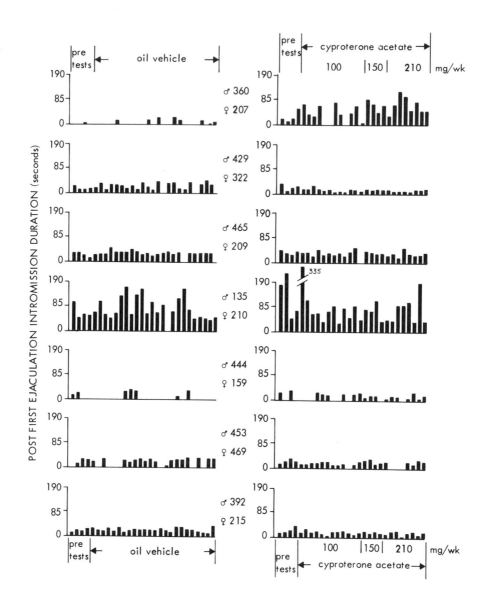

Figure 4. Time (in sec) a pair maintained a penile intromission following the first ejaculation per test observed in seven pairs stumptailed macaques during 12 weeks with CA injections in three dosages and during 12 weeks with oil vehicle injections. Note the inter-pair differences and intra-pair consistencies.

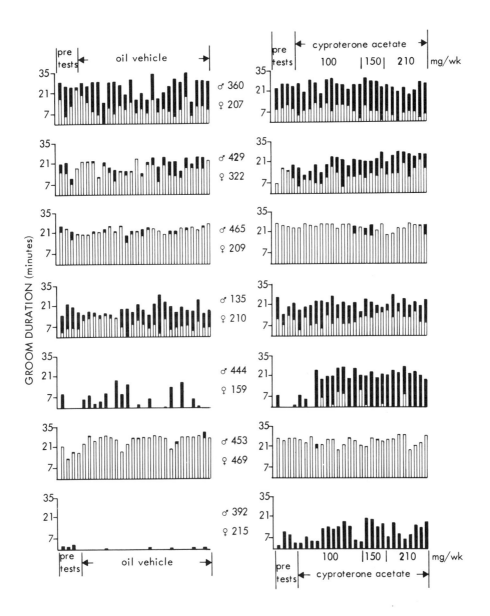

Figure 5. Time (in min) a female (open columns) and a male (black columns) spent grooming the partner per test of seven pairs of stumptailed macaques during 12 weeks with CA injections in three dosages and during 12 weeks with oil vehicle injections. Note the inter-pair differences and intra-pair consistencies.

MALE		PRE-TREATMENT	2 WEEKS 100 mg CA/wk	2 WEEKS 150 mg CA/wk	2 WEEKS 210 mg CA/wk	1 WEEK AFTER LAST CA ADMINISTRATION
ARY (inject.)	Testosterone	18.1	6.4	3.3	4.2	10.6
	Cyproterone acetate	N.D.	120	320	570	50
	Testislength	51	49	48	46	44
BILLY (tablet)	Testosterone	9.1	5.6	1.6	2.2	2.4
	Cyproterone acetate	N.D.	40	90	240	N.D.
	Testislength	46	42	41	40	39

Table 6. Serum testosterone and cyproterone acetate concentrations (ng/ml), as well as mean testislength (mm) of two male stumptailed macaques before, during and just after administartion of cyproterone acetate.
Male 'Ary' received CA dissoloved in oil (2 or 3 injections per week); male 'Billy' received CA as a tablet (Androcur®; 2 or 3 times per week).

five out of seven pairs during CA treatment. Secondly, pairs had specific patterns of grooming. In pair 392-215 (and to a lesser extent pair 444-159) only the male groomed, in contrast to pair 453-469 (and to a lesser extent pair 465-209) where only the female groomed. In the other three pairs both animals groomed, either simultaneously or alternating.

SOME ADDITIONAL DATA

During the course of these experiments we obtained two laboratory born young adult male stumptailed monkeys: Ary and Billy; body weights around 10 kg. These animals were housed individually and were initially not subjected to behavior tests. They were used in order to get some additional pharmacological/endocrinological information.

Between November 1979 and January 1980 both animals received CA: Ary by intramuscular injections and Billy in an oral tablet (Androcur^R). The

	WEEKS FOLLOWING START BEHAVIOR TESTS			1 WEEK AFTER LAST CA INJECTION AND BEHAVIOR TEST
	1	3	6	
	NO CA TREATMENT	2 WEEKS 210 mg CA/wk	3 WEEKS 300 mg CA/wk	
Body weight (kg)	8.4	9.8	8.8	9.0
Testislength (mm)	40	38	39	39
Testisvolume (ml)	18	16	16	14
Testosterone (ng/ml)	17.2	1.8	6.7	2.5
Cortisol (ng/100ml)	17.2	19.2	12.4	-
Mount	7.5	8.0	4.7	
Intromission	4.5	4.0	3.8	
Ejaculation	2.5	3.2	3.2	
Latency to 1st ejac. (min)	2.4	1.3	0.6	
Post 1st ejac. introm. (min)	1.4	1.5	0.8	
Yawn	12.5	6.0	2.2	
Male groom (min)	6.6	10.1	6.3	
Female groom (min)	11.1	6.0	8.9	

Table 7. Various physical, hormonal and behavioral data (Mean per test) of male stumptail 'Billy' before, during and just after daily oral administration of cyproterone acetate (Androcur®) tablets. The female partner was 327; behavior tests: 30 min, twice per week.

tablet was mixed with favorite food. For the treatment dose see Table 6. At regular intervals the animals were lightly anesthetized, a 5 ml blood sample was collected and the same physical measures were taken as described before.

From Table 6 it is clear that both methods of CA administration were effective in reducing testis size and serum testosterone levels, although the oral administration seemed to be more effective in lowering the serum testosterone.

In the light of these findings we hypothesized that the lack of a reduction in male sexual behavior in the present experiments I, II, and III (in contrast to the earlier mentioned clinical experiments, where CA was usually administered as a tablet) could be caused by the way CA was administered (i.e., dissolved in oil and injected i.m.). Therefore, an additional experiment was carried out.

Between June and August 1980 one male (Billy) was subjected to 30 minutes heterosexual behavioral pair-tests (with female 327), while receiving daily CA tablets (Androcur^R) in two dosages: 2 weeks 210 mg/wk, followed by 3 weeks of 300 mg/wk.

As can be seen from Table 7 serum testosterone levels and testis size decreased by CA, but copulatory activity was increased. From these results it was concluded that the absence of a reducing effect of cyproterone acetate on male sexual behavior in the stumptailed monkey could not be explained by the way the drug was administered.

GENERAL DISCUSSION

The present investigation has clearly shown that cyproterone acetate administration to male stumptailed monkeys does not reduce their heterosexual activity. On the contrary, long term CA administration by injection raised the frequency of copulation. The sexual motivation ("libido"), as measured by the latencies to first mount and to first ejaculation, was not affected by CA treatment. Moreover, cyproterone acetate treatment appeared to make the male monkeys more social as indicated by an increase in the time they spent grooming their female partner.

In light of the animal literature these findings are not surprising (see Introduction), but compared to the human literature there seem to be discrepancies. Many clinical investigators report a reduction in libido and potency in men treated with cyproterone acetate (e.g., Morse et al., 1973; Davies, 1974, 1975; Laschet and Laschet, 1975; Hoffet, 1980). However, most of these studies were badly designed, with no proper controls and were usually carried out on patients or 'deviants'. Still, the few available controlled clinical studies reported virtually the same inhibiting effects of CA on male sexual motivation and sexual behavior (e.g., Binder et al., 1971; Bancroft et al., 1974; Murray et al., 1975; Cooper, 1981).

Although there were no inhibiting effects on male stumptail sexual behavior in the present study, the decrease in serum testosterone levels, as well as the decrease in testicular size, clearly indicate that CA was physiologically active in these animals. The decrease in serum testosterone was comparable to what has been reported in the human male (Morse et al., 1973; Murray et al., 1975; Bartsch et al., 1977; van Wayjen and van de Ende, 1980; Cooper, 1981).

In clinical studies cyproterone acetate is usually administered in tablets in a dose of 100 to 200 mg/day. Therefore, one could hypothesize that the i.m. injection method used in the present investigation was less effective. This possibility seems very unlikely since it has been reported that injections of 300 to 600 mg CA at intervals of seven to 14 days achieved the same therapeutic (e.g., behavioral) effects as the oral administration of 100 or 200 mg daily (Laschet and Laschet, 1975). In the present study (Experiment III) CA was injected to male monkeys (body weight about 9 kg) for 12 weeks in dosages increasing from 100 mg to 210 mg per week. Therefore, we can say that in the present monkey study CA was administered long enough and in dose high enough to cause, compared to human data, a decrease in sexual behavior. The question remains, then, why was this inhibitory effect of cyproterone acetate not found in the male stumptailed macaque?

The answer might very well be that the male stumptailed monkey is not very dependent on gonadal steriods for the expression of sexual behavior. Support for this hypothesis can be found in the preliminary results of an ongoing castration study. Until 12 weeks following castration there is no striking reduction in sexual behavior of three male stumptailed macaques (Schenck and Slob, unpublished). It should also be noted that the serum testosterone levels during long term CA administration in the present study never dropped to extremely low levels. This finding corroborates clinical data (e.g., Morse et al., 1973; Bartsch et al., 1977; Cooper, 1981). Therefore, it is difficult to comprehend that in the human male CA treatment should decrease libido and potency through its hormonal action.

In the human male many investigators assume a direct causal relationship between serum testosterone (or its derivatives) and libido and potency

("...(Androcur) is used clinically for treatment of various androgen-dependent disorders such as hypersexuality...," Panesar et al., 1979). These investigators subsequently assumed that the decrease in serum testosterone by CA treatment is responsible for the decrease in male sexual behavior (see review by Hoffet, 1980). However, the relationship between human sexual behavior and serum testosterone is far from clear and much in debate (e.g., Beach, 1961, 1965; Swyer, 1968; Herbert, 1977). For example, reports on the effects of castration in adult men vary from "...retention of sexual capacities and responsiveness is ostensibly complete" to "...decrease or total loss" (Money, 1961, p. 1387). It may be noted in passing that similar variability has been reported for the effects of castration on male sexual behavior in the rhesus monkey (Phoenix et al., 1973).

It seems unlikely that CA affects male sexual behavior through its antiandrogenic and/or antigonadotrophic properties. Support for this statement can be found in the findings of clinical studies which reported that CA was effective within 1 or 2 weeks after the start of the treatment (Morse et al., 1973; Laschet and Laschet, 1975), as well as reports about a very fast reappearance in libido and potency following cessation of CA administration (from 1 day up to 1 or 2 weeks, Morse et al., 1973).

Another mode of action for cyproterone acetate on human sexual behavior has been proposed by Itil and collaborators. These investigators studied somatosensory evoked potentials (SEP) and computer EEG profiles (CEEG) of physically and mentally healthy male subjects before and after the administration of placebo, cyproterone acetate and mesterolone injection(s) (Itil et al., 1974; Saletu et al., 1975). From results of their CEEG experiments these authors suggested that cyproterone acetate "...may well have demonstrable CNS effect independent of its known hormonal effects" (Itil et al., 1974, p. 1167). In the discussion on their SEP paper they reach the following conclusion: "It seems to us that the above described improvement in psychopathology might be due to a direct anxyolitic-tranquilizing effect of cyproterone acetate"

(Saletu et al., 1975, p. 1326). The results of Cooper's study also pointed to the tranquilizing actions of CA. "Thus, the majority of patients noted that they were less irritable, more tolerant, and more relaxed physically and mentally during its administration" (Cooper, 1981).

Support for the hypothesis that CA mainly affects human male sexual behavior through an anxyolitic-tranquilizing action, might also be found in the present investigation. During cyproterone acetate administration the laboratory housed male stumptailed monkey showed an increase in heterosexual behavior as well as an increase in a very social behavior (Goosen, 1980): grooming the female partner.

ACKNOWLEDGEMENTS

This research was financially supported by the Netherlands Psychonomics Foundation, which is subsidized by the Netherlands Organization for the Advancement of Pure Research (ZWO).

We greatly acknowledge the skillful help of H. Koning and animals caretakers at the Primate Center TNO, Rijswijk, The Netherlands, where this investigation was carried out. We thank Dr. J.T.M. Vreeburg and M.J. Ooms for performing the testosterone assays, and T. Huizer and J. Pisa-Minderman for their help with the behavior tests. The support and help of Mr. A. Houba, director Schering Nederland BV is gratefully acknowledged. Thanks are also due to Professor J.J. van der Werff ten Bosch for critical reading and discussing the manuscript, as well as for his continuous interest and support of this research.

Part of these results have been presented as papers read at the Eastern Conference on Reproductive Behavior (New Orleans, USA, 1979) and at the Dutch Federation Meeting (Nijmegen, The Netherlands, 1980). A preliminary description of some of the results has already been published in abstract form (Slob et al., 1980). Part of the data have recently been published (Slob and Schenck, 1981).

REFERENCES

Adkins, E.K. and Mason, P. Effects of cyproterone acetate in the male Japanese quail. Horm. Behav. 5, 1-6 (1974).

Ågmo, A. Cyproterone acetate diminishes sexual activity in male rabbits. J. Reprod. Fert. 44, 69-75 (1975).

Abrahamsson, L., Hackl, H., Sogn, J., and Stafsnes, H. Klinische Erfahrungen und hormonelle Aspekte bei der Behandlung von Virilisierungserscheinungen bei der Frau mit Cyproteronazetat. Wiener klin. Wochenschr. 91, 126-130 (1979).

Bancroft, J., Tennent, G., Loucas, K., and Cass, J. The control of deviant sexual behaviour by drugs: I. Behavioural changes following oestrogens and anti-androgens. Brit. J. Psychiat. 125, 310-315 (1974).

Bartsch, W., Horst, H.-J., Becker, H., and Nehse, G. Sex hormone binding globulin binding capacity, testosterone, 5-dihydrotestosterone, oestradiol and prolactin in plasma of patients with prostatic carcinoma under various types of hormonal treatment. Acta Endocr. 85, 650-664 (1977).

Baum, M.J., Slob, A.K., de Jong, F.H., and Westbroek, D.L. Persistence of sexual behavior in ovariectomized stumptail macaques following dexamethasone treatment or adrenalectomy. Horm. Behav, 11, 323-347 (1978).

Beach, F.A. Hormones and Behavior. Cooper Square Publ., New York (1961).

Beach, F.A. Sex and Behavior. J. Wiley and Sons, New York (1965).

Binder, S., Roters, G., and Schultka, H. Klinische und experimentell-psychologische Untersuchungen zur Wirkung des Antiandrogens Cyproteronacetat bei erheblich vermindert zurechnungsfähigen und zurechnungsfähigen Sexualdelinquenten. Nervenarzt 42, 26-32 (1971).

Block, G.J. and Davidson, J.M. Behavioral and somatic responses to the antiandrogen cyproterone. Horm. Behav. 2, 11-25 (1971).

Cooper, A.J. A placebo-controlled trial of the antiandrogen cyproterone acetate in deviant hypersexuality. Compreh. Psychiat. (in press).

Davies, T.S. Cyproterone acetate for male hypersexuality. J. Int. Med. Res. 2, 159-163 (1974).

Davies, T.S. Collaborative clinical experience with cyproterone acetate. J. Int. Med. Res. 3, suppl. 4, 16-19 (1975).

Goldfoot, D.A., Slob, A.K., Scheffler, G., Robinson, J.A., Wiegand, S.J., and Cords, J. Multiple ejaculations during prolonged sexual tests and lack of resultant serum testosterone increases in male stumptail macaques (Macaca arctoides). Arch. Sex. Behav. 4, 547-560 (1975).

Goosen, C. On Grooming in Old Word Monkeys. Ph.D. thesis, University of Leiden, The Netherlands (1980).

Hammerstein, J., Meckies, J., Leo-Rossberg, I., Moltz, L., and Zielske, F. Use of cyproterone acetate (CPA) in the treatment of acne, hirsutism and virilism. J. Ster. Biochem. 6, 827-836 (1975).

Hammerstein, J., Lachnit-Fixson, U., Neumann, F., and Plewig, G. (eds.) Androgenization in Women. Excerpta Medica, Amsterdam (1980).

Herbert, J. The neuroendocrine basis of sexual behavior in primates, in Handbook of Sexology, J. Money and H. Musaph, eds. Elsevier/North-Holland Biomedical Press, New York (1977).

Hoffet, H. Die klinische Anwendung von Antiandrogenen in der Psychiatrie. Gynäkologie 13, 33-43 (1980).

Horn, H.J. Administration of antiandrogens in hypersexuality and sexual deviations, in Handbook of Experimental Pharmacology, Vol. XXXV/2, O. Eichler, A. Farah, H. Herken, and A.D. Welch, eds. Springer Verlag, Berlin (1974).

Horn, H.J. Role of antiandrogens in psychiatry, in Androgens and Antiandrogens, L. Martini and M. Motta, eds. Raven Press, New York (1977).

Itil, T.M., Cora, R., Akpinar, S., Herrmann, W.M., and Patterson, C.J. "Psychotropic" action of sex hormones: computerized EEG in establishing the immediate CNS effects of steroid hormones. Current Therap. Res. 16, 1147-1170 (1974).

Jacobi, G.H., Altwein, J.E., Kurth, K.H., Basting, R., and Hohenfellner, R. Treatment of advanced prostatic cancer with parenteral cyproterone-acetate: a phase-III randomised trial. Brit. J. Urol. 52, 208-215 (1980).

de Jong, F.H. and van der Molen, H.J. Determination of dehydroepiandrosterone and dehydroandrosterone sulphate in human plasma using electron capture detection of 4-androstene-3,6,17-trione after gas-liquid chromatography. J. Endocrin. 53, 461-474 (1972).

Kirk, R.E. Experimental Design: Procedures for the Behavioral Sciences. Brooks/Cole, Washington (1968).

Laschet, U. and Laschet, L. Antiandrogens in the treatment of sexual deviations of men. J. Ster. Bioch. 6, 821-826 (1975).

Michael, R.P., Plant, T.M., and Wilson, M.I. Preliminary studies on the effects of cyproterone acetate on sexual activity and testicular function in adult male rhesus monkeys (Macaca mulatta), in Advances in the Biosciences 10, G. Raspé, ed. Pergamon Press, Vieweg, Oxford (1973).

Money, J. Sex hormones and other variables in human erotocism, in Sex and Internal Secretions, II (3rd ed.), W.C. Young, ed. Williams & Wilkins, Baltimore (1961).

Morse, H.C., Leach, D.R., Rowley, M.J., and Heller, C.G. Effect of cyproterone acetate on sperm concentration, seminal fluid volume, testicular cytology and levels of plasma and urinary ICSH, FSH and testosterone in normal men. J. Reprod. Fert. 32, 365-378 (1973).

Murray, M.A.F., Bancroft, J.H.J., Anderson, D.C., Tennent, T.G., and Carr, P.J. Encodrine changes in male sexual deviants after treatment with anti-androgens, oestrogens or tranquilizers. J. Endocr. 67, 179-188 (1975).

Neumann, F. and Steinbeck, H. Antiandrogens, in Androgens II and Antiandrogens. Springer Verlag, Berlin (1974).

Neumann, F., Gräf, K.-J., Hasan, S.H., Schenck, B., and Steinbeck, H. Central actions of antiandrogens, in Androgens and Antiandrogens, L. Martini and M. Motta, eds. Raven Press, New York (1977).

Neumann, F., Schleusener, A., and Albring, M. Pharmacology of antiandrogens, in Androgenization in Women, J. Hammerstein, U. Lachnit-Fixson, F. Neumann, and G. Plewig, eds. Excerpta Medica, Amsterdam (1980).

Nieuweboer, B. and Lübke, K. Radioimmunological determination of cyproterone acetate. Hormone Res. 8, 210-218 (1977).

Panesar, N.S., Herries, D.G., and Stitch, S.R. Effects of cyproterone and cyproterone acetate on the adrenal gland in the rat: studies in vivo and in vitro. J. Endocr. 80, 229-238 (1979).

Phoenix, C.H., Slob, A.K., and Goy, R.W. Effects of castration and replacement therapy on sexual behavior of adult male rhesus. J. Comp. Physiol. Psychol. 81, 472-481 (1973).

Plant, T.M., and Michael, R.P. Testicular function of adult male rhesus monkeys. J. Endocr. 57, xli-xlii (1973).

Saletu, B., Saletu, M., Herrmann, W.M., and Itil, T.M. Are hormones psychoactive? Evoked potential investigations in man. Arzneim.-Forsch. (Drug Res.) 25, 1321-1327 (1975).

Schmidtke, D., and Schmidtke, H.-O. Ein neues Antiandrogen beim Hund. Kleintier-Praxis 13, 146-149 (1968).

Slob, A.K. and Nieuwenhuijsen, H. Heterosexual interactions of pairs of laboratory-housed stumptail macaques (Macaca arctoides) under continuous observation with closed-circuit video recording. Int. J. Primat. 1, 63-80 (1980).

Slob, A.K. and Schenck, P.E. Chemical castration with cyproterone acetate (Androcur^R) and sexual behavior in the laboratory-housed male stumptailed macaque (Macaca arctoides). Physiol. Behav. 27, 629-636 (1981).

Slob, A.K., Baum, M.J., and Schenck, P.E. Effects of the menstrual cycle, social grouping, and exogenous progesterone on heterosexual interaction in laboratory housed stumptail macaques (M. arctoides). Physiol. Behav. 21, 915-921 (1978a).

Slob, A.K., Wiegand, S.J., Goy, R.W., and Robinson, J.A. Heterosexual interactions in laboratory-housed stumptail macaques (Macaca arctoides): observations during the menstrual cycle and after ovariectomy. Horm. Behav. 10, 193-211 (1978b).

Slob, A.K., Ooms, M.P., and Vreeburg, J.T.M. Annual changes in serum testosterone in laboratory housed male stumptail macaques (M. arctoides). Biol. Reprod. 20, 981-984 (1979).

Slob, A.K., Schenck, P.E., and Nieuwenhuijsen, H. The effects of cyproterone acetate on social and sexual behavior in the adult male laboratory housed stumptail macaque (M. arctoides). Antropol. Contempor. 3, 272-273 (abstract) (1980).

Smals, A.G.H., Kloppenborg, P.W.C., Goverde, H.J.M., and Benraad, T.J. The effect of cyproterone acetate on the pituitary-adrenal axis in hirsute women. Acta Endocr. 87, 352-358 (1978).

Swyer, G.I.M. Clinical effects of agents affecting fertility, in Endocrinology and Human Behaviour, R.P. Michael, ed. Oxford University Press, London (1968).

Underhill, R. and Dewhurst, J. Further clinical experience in the treatment of hirsutism with cyproterone acetate. Br. J. Obst. Gyn. 86, 139-141 (1979).

Verjans, H.L., Cooke, B.A., de Jong, F.H., de Jong, C.M.M., and van der Molen, H.J. Evaluation of a radioimmunoassay for testosterone estimation. J. Ster. Biochem. 4, 665-676 (1973).

Whalen, R.E. and Edwards, D.A. Effects of the anti-androgen cyproterone acetate on mating behavior and seminal vesicle tissue in male rats. Endocrinology 84, 155-156 (1969).

Wayjen, R.G.A. van and van den Ende, A. Metabolic effects of cyproterone acetate, in Androgenization in Women, J. Hammerstein, U. Lachnit-Fixson, F. Neumann, and G. Plewig, eds. Excerpta Medica, Amsterdam (1980).

Werff ten Bosch, J.J. van der. Manipulation of sexual behaviour by anti-androgens. Psychiatry, Neurologia, Neurochirurgia 76, 147-149 (1973).

Zucker, I. Effects of an anti-androgen on the mating behaviour of male guinea-pigs and rats. J. Endocr. 35, 209-210.

Chapter 8
Effects of Methaqualone on Social-Sexual Behavior in *Macaca Mulatta*

G. Claus and A. Kling

INTRODUCTION

During the past few years we have been actively involved in the investigation of the effects of certain psychotropic drugs on social behaviors of rhesus monkeys (Macaca mulatta) in a colony setting. Parallel to these studies, we have conducted multi-channel telemetered EEG experiments on the same species, and attempts have been made to correlate the behavioral changes observed under the effects of drugs with the EEG patterns obtained simltaneously from different regions of the brain.

In the present paper, we shall deal only with the results of those experiments which utilize methaqualone. Methaqualone is a quinazolinone derivative and serves as an hypnotic sedative agent which is unrelated to any other kind of sedative or to the major or minor tranquilizers. It was first synthesized by Kackher and Zaheer (1951), and good results were obtained with its applications as an hypnotic, even in cases of intractable insomnia (Deibez and Graner, 1967). Unlike barbiturates, the drug does not cause hangover, and has less addictive potential (Sargant, 1973; Kohli et al., 1974). In the early 1970's it became a favorite of the street drug culture under the name of "love drug." The Food and Drug Administration has classified the compound as addictive and included it in Schedule II.

There are a few anecdotal studies in the literature, based mainly on accounts of users, in which it is both claimed and denied that the drug has aphrodisiac potency in group settings (Gamage and Zerkin, 1973; Gerald and Schwirian, 1973; Inaba, et al., 1973; Lewis and Steindler, 1973). More acute observations regarding the purported effect of the drug may be found in Gelpke (1975) who hypothesized that methaqualone is capable of releasing and fusing the potential of the users for creative activity. According to this author, if a group does not have an appropriate level of intelligence, the release of creativity will degenerate into orgies; whereas among intellectuals, joint works of art may be produced by the participants.

Behavioral changes in humans under the influence of many other drugs favored by abusers for their mood-altering potential have been somewhat more thoroughly reported than they have for methaqualone. However, even in the case of alcohol, where a vast literature of studies dealing with individual and group behavior under the influence of ethanol intoxication has grown up over recent decades, interpretations regarding the relative influence of the pharmacology of the drug versus social influences vary widely and are still in flux. For example, the notion that ethanol, at least in early phases of intoxication, has a disinhibiting effect on certain behaviors -- specifically aggressivity and sexual arousal -- has been thrown into doubt by the thorough studies of MacAndrew and Edgerton (1969), who showed that social factors are the major determinants of whether or not violence will occur after drinking and that the pharmacological action of ethanol is minimal in this context.

Consideration of social expectations regarding the effects of particular drugs in differing societies, or within subcultures, may help to clarify the relationship between drug effects per se and socially conditioned responses. However, it may be argued that observations of human behavioral changes under the influence of drugs could only be separated from the factor of social expectation by utilizing completely naive volunteers -- a method which raises a number of ethical issues, especially if the drugs involved have addictive potential, or the behaviors of interest to the investigators touch on social mores. In this connection, Crowley and his co-workers (1974), who were interested in testing the effects of certain widely abused drugs on aggressivity and sexual behaviors, proposed that the practical and ethical obstacles to administering drugs in a laboratory setting to naive human volunteers could be obviated by utilizing a monkey colony, since these animals live in groups with rich varieties of social behaviors.

While findings from primate experiments obviously cannot be uncritically generalized to humans, the model does seem to provide a number of advantages for the study of drugs of abuse: freedom from social set; freedom to observe individual and group behaviors in a quasi-natural environment; and the possibility of administering potentially dangerous doses.

The exact site of action of methaqualone in the brain in unknown, but claims have been advanced that it must influence different regions from either the barbiturates or glutethimide (Soulairac and Gottesmann, 1967; Bhargava et al., 1972). In a first attempt to investigate whether a specificity of response could be elicited in the monkeys, which would be both qualitatively and quantitatively different from that obtained with other drugs previously studied (Cressman and Cadell, 1971; Kjellberg and Randrup, 1972; Redmond et al., 1971a; Redmond et al., 1971b; Miller et al., 1973; Crowly et al., 1974; Miller and Geiger, 1976), experiments were undertaken in which repeated doses of methaqualone were given to three target animals in an established monkey colony of ten (Claus et al., 1980). In this phase of the study, we were interested primarily in observing the effects of the drug on the dosed animal, and noting whether or not the behavior of subjects would alter the behavior of other colony members. Eight specific questions were raised in this connection:

1. Would the dosed animal show any new behaviors, not present during the baseline studies?
2. Would any of the usual behaviors of the animal disappear as a consequence of the drugged state?
3. Would any quantitative change occur in specific behaviors, such as an increase or decrease in aggressivity?
4. Would social rank of the animal be affected; and if so, would this change be reversed after the drug wore off?

5. Would the drug have any prolonged effects on the behaviors of the treated animals?
6. Would there be sex differences?
7. Would any learning occur during the dosed state which would permit the animal to better cope with subsequent drug-induced behaviors?
8. Would the altered behaviors of the treated animal give rise to changes in the activities of other animal(s) in the colony?

In the next experiments, we were interested in exploring what would be the effects on social behaviors when several animals were dosed simultaneously, since among human users the drug is seldom taken by a single individual (Claus et al., in press). In effect, the anecdotal claims regarding the aphrodisiac potency of this drug frequently refer to group settings. We also wanted to clarify whether or not sex differences might exist in the responses of the animals, since in the first experimental series only one female was tested.

MATERIALS AND METHODS

Subjects

In the first experimental series, a well established social group (two years), composed of ten macaques (<u>Macaca mulatta</u>) was used: four adults (one male and three females), three subadult females, one adolescent male, and two new-borns. The behavior of the infants was not included in the observations. The rank order of the animals was determined by quantitative observations of access to food, cage position, and displacements.

In the second experimental series, three animals were caged together, forming a small colony. It consisted of an adult male, who had been caged separately from infancy and had had no social experience. The second member was an adult female; while the third one was a subadult female. Earlier, the two female animals were members of a large colony and had had extensive social experience.

For the EEG studies, two adult males were used, who had been operated upon for the placement of electrodes in selected structures of both cortical and subcortical regions.

Apparatus

For the first experimental series, observations were conducted on the colony as a whole, while one of its members was under the influence of the drug. The colony cage was six meters long, four meters wide, two-and-a-half meters high, and had three sides of sealed concrete block walls. Its fourth side was constructed of a wire fence, as was its ceiling. On the fence in the front, there was an entrance, permitting the capture of individual animals and the cleaning of the cage. On the same fence, a guillotine door was located, to which an individual cage could be attached. After a monkey had been chased or lured into the small cage, the door could be lowered and the animal could be temporarily removed from the colony. In front of the wire mesh there was a space approximately one-and-one-half meters wide, where cleaning equipment and food buckets were stored. The entire room was also enclosed from the front by a cinder block wall, which had a door to the outside corridor. On the two side walls, one-way mirrors were installed, and the observations were conducted from behind a curtain in the neighboring room. An automatic water dispenser was used with a standard canine "Lixit" valve. Perches made of two-inch galvanized pipe were attached to the back wall, to which a climbing ladder was added. The room was temperature controlled and the lights were automatically turned off in the evening and came on in the morning, permitting a twelve hour light period.

The experiments where all members of the small colony were injected and simultaneously observed were conducted in an adjacent room. In this room there was a chamber (2 x 2 x 3 meters) constructed of Lexan on all four sides and with bars of fiberglass. On one side there was a guillotine door for introducing or removing animals from the chamber.

The cage was adjacent to a door with a built-in one-way mirror; and the observations were conducted from the room next to the chamber.

For the telemetered EEG experiments, the same chamber was used, and the instruments were located in the adjacent room. The instrumentation consisted of a polygraph recorder for the continuous tracing of the electrical activities originating from the brain of the operated animal. These were simultaneously put on magnetic tape for computer analysis. Details of the six-electrode time-multiplex telemetry unit have been published (Deutsch, 1979). Briefly, RF to AF feedback was prevented by mode switching at the basic cycle rate, 400 Hz, so that the RF transmission was cut off while the AF amplifiers were on, and vice versa. The channel rate was 12,800 Hx. The channel six input consisted of 200 Hx reference signal. The unit weighed 20 grams and drew 10 mA at 5V. Each channel band width extended from).36 to 59 Hz. The output was FM with a peak deviation of 75 kHz of a 20 MHz carrier. The decoder locked on a 400 Hz AM sync signal. RF carrier shift due to motion artifact was fed back to the data channels so as to cancel out their shift components. A microphone attached to the instrument permitted the dictation of the animal's activity in the cage to the tape; thus when the quantitative EEG analysis was performed, it was possible to ascertain whether or not the test animal was walking, sitting, climbing, etc. These activities were also marked down on the graphed EEG recording.

An analysis was performed on a PDP-11/04 computer which was equipped with eight analog inputs. This multichannel analysis was performed in two stages: first, frequency spectra were established, followed by statistical computations (means, standard deviations, and correlations). The results were plotted on an IBM 1620 incremental digital plotter.

The computer sampled a five-second long EEG from the FM tape recorder of a particular episode. The input to the computer was band-limited to about 40 Hz, with a sampling rate of 128 samples/second. After sampling was completed, the five-second long record was divided in five one-second long segments, each having sample points. The computer then performed a Fast Fourier Transfer (FFT) on each one-second segment.

The computer took an average of the five FFT's and this averaged frequency spectrum was stored on a floppy disc for subsequent analysis and plotting. Averaging of FFTs had proved useful in correcting for noise.

Materials

Pure methaqualone base was obtained from Arnar-Stone Laboratories, Inc. and prepared for injection according to the method of Saxena et al., (1972). It was dissolved in boiling polyethelene glycol (PEG-200), and the solution was pipetted into serum vials, autoclaved at 120 degrees C. at 15 lbs. for two hours. Prior to the experiments, a dose-response curve was established on a separately caged animal which was injected with 2.5, 5, 7.5, 10, 12.5, and 15 mg/kg methaqualone. 10 mg/kg was selected for the experiments, based on the fact that at this dose level, the test animal became slightly ataxic but did not fall asleep.

Experimental Design

In the first series, the focal sampling method was used for the observation of behaviors: watching the activities of each animal for twelve minutes, in a random order, with the exception of the babies. A total of twenty hours was recorded in this manner to obtain a baseline, which yielded data approximating two hours per animal.

Three animals were selected for the experiments: the dominant male, mid-ranking female, and the adolescent male. The experiments were designed as follows. The target animal was injected with a weight-adjusted dose of methaqualone (10 mg/kg), and his or her behavior was observed and recorded continuously for a two-hour period. At the same time, 12-minute observations were carried out on each of the rest of the colony members. A few days following the drug experiment, an equal amount of saline was given to the same animal as a control. This procedure was repeated three times, care being taken that no two drug experiments should occur in the same week, in order to avoid possible habituation. During the saline controls, the target animals were again observed continuously for two hours and the rest of the colony members were monitored for 12 minutes each. A total of 43 active and passive behaviors were noted, and those which lasted longer than 30 seconds were named and counted, and later these were assigned an arbitrary duration of 20 seconds. These behaviors were termed brief interactions.

Means and standard deviations were calculated for the drug observations and the controls on all behaviors which seemd to show notable change; and those changes where no overlap was seen were regarded as significant and attributable to drug effect.

In the experiments with the small colony, where all the participating animals were under the influence of the drug, basically the same approach was used as outlined above, with the difference that each member's behavior was continuously and simultaneously recorded for two-hour periods. Six hours of baseline studies, ten hours of experimental observations (five experiments) and four hours of saline controls were conducted. The controls were initiated after the second and fourth methaqualone experiments.

For the EEG telemetry studies, the subjects were operated upon under intravenous pentobarbital sodium anesthesia. Electrodes were placed symmetrically, using stereotaxic technique, into the selected brain regions. The electrodes were then plugged into a multichannel radio transmitter, which was screwed into the skull. Its battery could be magnetically turned on or off in order to extend its useful lifetime. After a one-week recovery period, the animal was moved into the small chamber constructed of Lexan, as described above. A circular antenna was placed above the chamber, and the animal's behavior was observed for about 20 minutes from behind the one-way mirror door, and his EEG recorded as described earlier. Next, the animal was injected with 10 mg/kg dose of methaqualone, and his EEG was recorded for two hours while being observed.

Behavioral Categories

The behavioral categories used in the present studies did not adhere strictly to the behavioral taxonomy of Kaufman and Rosenblum (1976) for macaques. In general, they fell into two broad areas: affiliative and nonaffiliative behaviors. The affiliative behaviors included such descriptors as sitting together, huddling, or active and passive grooming. The nonaffiliative behaviors comprised pacing, walking, climbing, or sitting alone. The brief interactions were again composed of two broad categories: aggressive (antagonistic) behaviors and "friendly" interactions, in the sense of Hinde (1972). Under the first, displacing, threatening, slapping, fighting, etc. were tabulated; while among the second, approaching, presenting, mounting, etc. were enumerated. Some behaviors which occurred only under the influence of the drug comprised acts of balancing, slipping, napping, or erotic arousal and autoerotic activities. These were either timed or simply noted as to their frequency. Each behavior has been described in detail and assigned a category in our previous publications (Claus et al., 1980; Claus et al., in press) to permit standardization.

The experiments were conducted under the observation of three independent persons, in order to rule out observer bias.

RESULTS

From the first experiments, for purposes of demonstration, we selected the results obtained on the dominant male and the mid-ranking female.

The nonaffiliative behaviors (See Note 1) of Big Daddy, the dominant male, are shown in Figure 1.

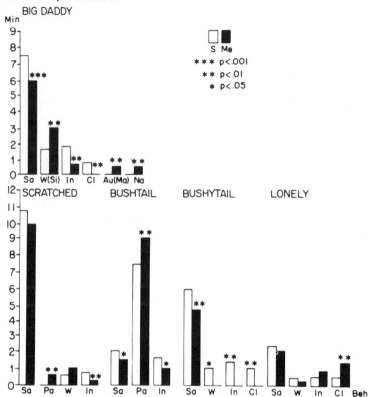

Figure 1. Nonaffiliative behaviors of Big Daddy and of other colony members while the former was the experimental subject. Graphs in Figs. 1-4 represent means of behaviors resulting from 12 hours of observation, interpolated to 12-minute time blocks and expressed on the ordinate as time spent engaged in the activities shown on the abscissa. White block (S) = saline; darkened block (me) = methaqualone. Only those behavioral changes are graphed which differed notably from baseline and saline findings.

Five new behaviors occurred, and there were also some changes in the durations of certain activities compared to the saline controls. This subject became involved in grooming others, and he also napped. When he was slightly ataxic, he slipped and performed balancing acts and later ceased climbing. In the second half of each drug treatment period, he performed autofellatio. Finally, he exhibited a general mood of indifference, during which periods the highest ranking female, Scratched, watched and kept order in the group.

On the same Figure 1 are also shown the behaviors of those subjects which changed notably while Big Daddy was dosed. The sitting alone behavior of Scratched, the highest ranking female, was decreased, while her pacing and walking was increased, an observation in accord with the fact that she took over the dominant role during those periods when Big Daddy was napping or otherwise inattentive. The mid-ranking female, Brushtail, apparently became markedly excited, as indicated by a considerable increase in her pacing activity. Bushy Tail, a subadult female, spent most of her time during these experiments in affiliative behaviors, which will be discussed later. Lonely, the adolescent male, was apparently more fearful and showed more climbing than usual.

Big Daddy's affiliative behaviors are shown in the next figure (Figure 2). As has been mentioned earlier, under the influence of the drug this subject groomed others. This activity apparently had effects on only the lower ranking subjects. The major change shown in Juvenile's affiliative behaviors from saline to drug is the fact that she was groomed. This does not have statistical significance on account of large standard deviations among the different saline control series. However, the change is important, because it represents a drug effect expressed as a role reversal. Normally, it would be Juvenile who would groom Big Daddy, whereas when Big Daddy was treated, she became the recipient of his grooming. Bushy Tail's affiliative behavioral changes were the most remarkable, with the increase in both active and passive grooming activities significant at the 0.001 level. During the experiments, she also spent considerably more time huddling (Hu). The increase in the affiliative behaviors of Lonely was worthy of note but not statistically significant. Lonely also spent more time in grooming and huddling activities.

Figure 3 presents the nonaffiliative behaviors while Brushtail was the target animal. In general, she became quieter, her sitting alone time

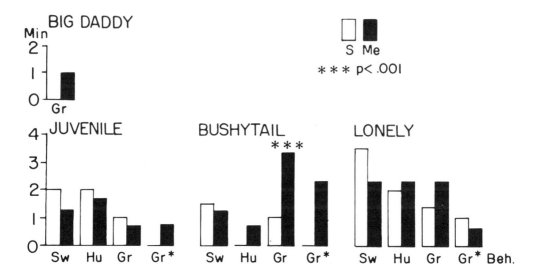

Figure 2. Affiliative behaviors of Big Daddy and of other colony members while the former was the experimental subject.

increased, and her pacing decreased considerably. Her generalized indifference is shown by the notable decrease in time spent inspecting. During the second half of the experiment, the animal adopted a new pattern. She became aggressive and initiated fights with older, higher ranking females. She seemed to be the best learner among the tested animals: whereas during the first two-hour experiment she slipped while pacing for a total of 36 minutes, in the next experiment she slipped only 17 minutes, and in the last a mere five minutes.

Comparing Figure 1 and Figure 3, one can see that some of Big Daddy's apparently significant behavioral changes during his treatment, such as the decrease in sitting alone time and the increase in walking, cannot be ascribed to the direct effect of the drug, because the same happened when Brushtail was dosed. On the other hand, the considerable

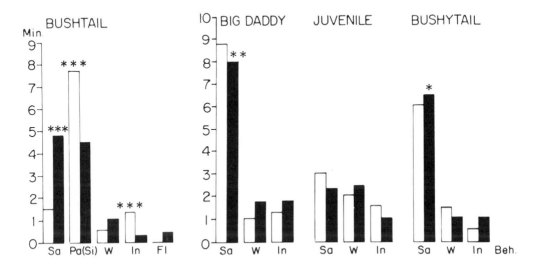

Figure 3. Nonaffiliative behaviors of Brushtail and of other colony members while the former was the experimental subject.

decrease in inspection when he was dosed is indeed ascribable to methaqualone, since his inspecting behavior increased when Brushtail was the target animal. The nonaffiliative behaviors of the other colony members did not change notably when Brushtail was treated except that Bushy Tail, a lower ranking female, showed some increase in climbing attributable to the target animals' increased aggressivity.

During treatment, Brushtail showed two new affiliative behaviors: she engaged in active grooming and also let herself be groomed (Figure 4). The most important changes among the other animals occurred between Jane, the lowest ranking female, and Lonely. They spent most of the time together, grooming. One might speculate that the aggressivity shown by Brushtail in the second half of the methaqualone experiment brought these two low ranking animals together.

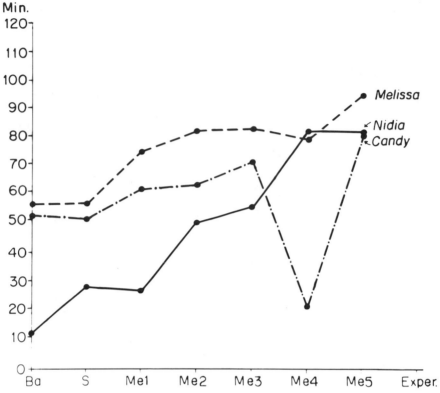

Figure 4. Affiliative behaviors of Brushtail and of other colony members while the former was the experimental subject.

Some comments should be made regarding changes observed in the brief interactions -- both friendly and antagonistic -- among colony members when Big Daddy and Brushtail were given methaqualone. Under the influence of the drug, Big Daddy's aggression against Scratched seemed to show a considerable increase. This, however, mainly resulted from displacement; when he awakened from his brief naps he again took over the dominant role, displacing Scratched from her position. His aggressivity against Brushtail, Juvenile, and Bushy Tail decreased under the influence of the drug; and he was even willing to tolerate some friendly approached from Brushtail, Bushy Tail, Jane and Lonely.

The brief interations of Brushtail when dosed were friendly in the first phase and aggressive in the second phase of treatment. The animal attacked

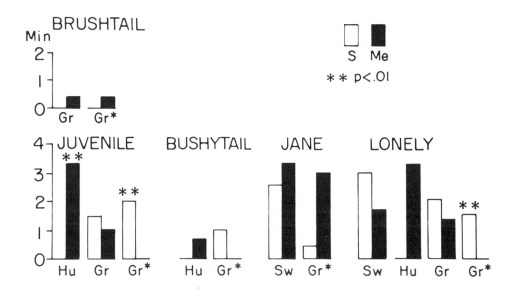

Figure 5. Total nonaffiliative behaviors of subject (timed). Note that Candy's behavior in Me4 is aberrant. This is the result of the fact that during this experiment, Nidia, the naive male, established his dominance.

Scratched and Wanda, the two highest ranking females. On the other hand, she was attacked by Juvenile and Bushy Tail, two females of lower rank, during the early phase of treatment. Early in the experiments, she showed some friendly approaches to Wanda, which normally she never exhibited; and she also approached Bushy Tail, another animal with whom she usually had no interaction. Despite her fighting episodes, Brushtail was approached in a friendly manner by Lonely. As the drug wore off, she became docile and groomed Scratched.

The results of the second experiments are shown in the next two figures. They represent changes in the total affiliative and nonaffiliative behaviors, respectively, of the three simultaneously dosed animals, including averages of the baseline and

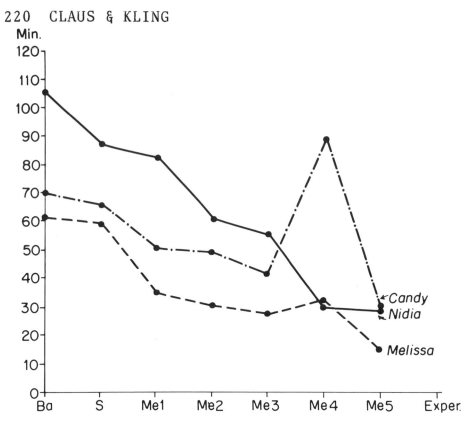

Figure 6. Total affiliative behaviors of subjects (timed). The two figures are close to mirror images of each other, the differences being due to the number of short interactions which are not graphed here. As above, again Candy's behavior in Experiment 4 seems to be aberrant.

saline results and findings from each of the methaqualone treatments. It is evident that over the course of the experiments all three animals show a gradual decline in affiliative behaviors paralleled by an increase in affiliative behaviors (Figures 5 and 6). There is one aberrant point for Candy (Experiment 4), which will be discussed later. The slight increase in total time spent by Nidia in affiliative behaviors from baseline to saline can be ascribed to adjustment on his part to colony interaction. It may also be seen that during the first methaqualone experiment, he still spent most of his time alone, whereas from the second experiment on, his interactions with one or both of the female animals increased markedly, reaching 83 minutes in the last two experiments.

Melissa was the dominant colony member in the baseline and saline studies. Her affiliative behaviors also increased under the influence of the drug, the increase being immediate with the first experiment, with less change over the course than occurred in the other animals. Candy interacted primarily with Melissa during baseline and saline observations, as well as in Experiment 1, whereas in the second, third, and fifth experiments her affiliative interactions reflected time spent with both of the two other animals separately, or, in the last methaqualone trial, with both together.

Table 1 compares the baseline studies and saline controls in terms of the time spent by each animal in specific behaviors observed. It is apparent that in spite of the interspersed methaqualone experiments, the majority of the animal's timed behaviors returned approximately to baseline levels when only saline was administered. Some degree of adjustment to being with other animals was shown by Nidia: decreases in his pacing and walking times paralleled increases in the time he sat with other animals or was groomed by them. However, when compared with the second through the fifth drug trials, his saline behaviors were still far closer to baseline observations. Thus, most of the changes in his behaviors during the methaqualone experiments can be ascribed to drug effect, with only minor increases in affiliative activities due to adaptation.

No animal showed any significant difference between baseline and saline studies in their brief interactions, friendly or antagonistic. It is especially noteworthy that Nidia made no mounts or attempted mounts during the saline trials, the second of which followed Experiment 4; whereas by Experiment 3 he had learned to mount Melissa properly, and during Experiment 4 he already copulated with her.

Table II shows the duration of behaviors which occurred only under the influence of the drug. It can be seen that in the first two methaqualone experiments, there was an increase in the aggressive behavior of Melissa towards Candy; while in the last two experiments it was the shy, previously isolated

TABLE I

Baseline Observation and Saline Controls of Subjects
(Numbers: Averages of Three and Two Experiments, Respectively)

Behaviors	Name of Animal/Duration of Behavior*					
Affiliative Behav. (Timed)	Nidia		Melissa		Candy	
	Baseline	Saline	Baseline	Saline	Baseline	Saline
Sitting with	4.0	14.0	9.8	9.5	10.0	11.0
Grooming	0.3	0.0	20.5	20.5	28.0	27.0
Groomed by	6.3	13.5	24.5	25.5	13.0	12.0
Nonaffiliative Behav. (Timed						
Pacing	70.6	62.5	8.8	8.0	0.0	0.0
Sitting alone	31.6	40.5	32.0	31.5	48.8	47.5
Walking	0.3	2.5	2.8	3.0	1.0	1.0
Inspecting	2.5	2.5	15.5	15.0	10.0	11.5
Climbing	0.0	0.0	1.5	2.5	5.2	6.0
Friendly, brief interactions (Time estimated)						
Approaching	1.6	1.3	1.6	0.6	0.6	0.6
Approached by	0.6	0.3	0.6	0.6	1.3	1.3
Presents	0.3	1.3	0.3	1.3	0.0	0.3
Presented by	1.0	1.0	0.0	1.3	0.3	1.6
Mounts	0.0	0.0	1.0	0.0	0.0	0.0
Mounted by	0.0	0.0	0.0	0.0	1.0	0.0
Antagonistic, brief interactions (Time estimated)						
Displaces	0.0	0.0	0.3	0.0	0.3	0.0
Displaced by	0.3	0.0	0.0	0.0	0.3	0.0

*Total time: 120 min.

male who showed aggression towards the subadult female, especially marked in Experiment 4. Over the course of the experiments, all animals learned to cope with drug effects which impaired their performance in one or another behavior. Nidia, who was an active "pacer", slipped for a total of 8.3 minutes in Experiment 1. By Experiment 3, this was reduced to 2.5 minutes, and in the last two experiments he did not slip at all. He also fell only during the first two experiments; while Candy fell in the first and third trials, but not later. The latter animal, who did a good deal of climbing under all conditions, reduced her unsuccessful climbs from 1.5 to 2.5 minutes in Experiments 1 and 2 respectively, and to zero in the last three experiments.

The most notable new behaviors were the sexual activities of the previously isolated and naive male. Even in Experiment 1, when he interacted very little with the two other animals, he showed a total of 7.5 minutes with erection. In contrast to our earlier experiments with male animals (Claus et al., 1980), where the drug was given to only one animal at a time in a colony setting, autoerotic behaviors such as masturbation and autofellatio were not seen in this series. On the other hand, in these drug trials the male showed erection while sitting with or being groomed by other animals; he also allowed himself to be masturbated by the adult female. True copulation was observed in the last two experiments. Finally, the male's total time with erection increased from 7.5 minutes in Experiment 1 to 16 minutes in the final experiment. It is apparent that it was not only the male who evidenced arousal under the influence of the drug and learned a spectrum of sexual interactions throughout the five experiments; but also that the females became inclined to seek close body contact and/or engage in sexual activities per se.

Methaqualone Experiment 4 requires special mention (cf. Figures 5 and 6). It shows several anomalies which seem in part to contradict the earlier generalizations. This is the only experiment, for example, in which active fighting occurred, in the sense of the original definition of pursuit, capture, and biting. Also, the sitting

TABLE II

New Behaviors Observed in Subjects During Methaqualone Experiments
(Numbers: Minutes per 2 hr. Drug Trials)

Name of Animal/Duration of Behavior*

Behaviors	Nidia					Melissa					Candy				
	Me 1	Me 2	Me 3	Me 4	Me 5	Me 1	Me 2	Me 3	Me 4	Me 5	Me 1	Me 2	Me 3	Me 4	Me 5
Affiliative Behav. (timed)															
Huddling	0.0	2.5	0.0	0.0	0.0	0.0	2.0	0.0	0.0	0.0	0.0	0.5	0.0	0.0	0.0
Masturbated by	0.0	0.0	5.0	0.0	0.0	0.0	0.0	0.0	0.0	0.0	0.0	0.0	0.0	0.0	0.0
Erection sitting with	1.5	0.0	4.0	1.0	10.0	-	-	-	-	-	-	-	-	-	-
Erection while groomed by	0.0	0.0	0.0	8.0	0.0	-	-	-	-	-	-	-	-	-	-
Nonaffiliative Behav. (timed)															
Slipping	8.3	4.3	2.5	0.0	0.0	0.6	0.0	0.0	0.0	0.0	0.0	0.0	0.0	0.0	0.0
Unsuccessful climb	0.0	0.0	0.0	0.0	0.0	0.0	0.0	0.0	0.0	0.0	1.5	2.5	0.0	0.0	0.0
Fall	1.0	1.3	0.0	0.0	0.0	0.0	0.0	0.0	0.0	0.0	0.3	0.0	0.6	0.0	0.0
Erection while sitting alone	3.5	5.0	4.0	3.0	4.5	-	-	-	-	-	-	-	-	-	-
Erection while pacing	2.5	1.0	1.0	0.5	0.5	-	-	-	-	-	-	-	-	-	-
Friendly, brief interactions (time estimated)															
Attempted mount	0.0	0.6	0.0	0.3	0.0	0.0	0.0	0.0	0.0	0.0	0.0	0.0	0.0	0.0	0.0
Attempted mount by	0.0	0.0	0.0	0.0	0.0	0.0	0.6	0.0	0.3	0.0	0.0	0.0	0.0	0.0	0.0
Copulation	0.0	0.0	0.0	0.6	1.0	0.0	0.0	0.0	0.6	1.0	0.0	0.0	0.0	0.0	0.0
Antagonistic, brief interactions (time estimated)															
Threatens	0.0	0.0	0.0	1.3	0.3	0.0	2.0	0.0	0.0	0.0	0.0	0.0	0.0	0.0	0.0
Threatened by	0.0	0.0	0.0	0.0	0.0	0.0	0.0	0.0	0.0	0.0	0.0	2.0	0.0	1.3	0.3
Slaps	0.0	0.0	0.0	0.6	0.3	0.0	0.0	0.0	0.0	0.0	0.0	0.0	0.0	0.0	0.0
Slapped by	0.0	0.0	0.0	0.0	0.0	0.0	0.0	0.0	0.0	0.0	0.0	0.0	0.0	0.6	0.3
Fight	0.0	0.0	0.0	3.0	0.0	0.3	0.3	0.0	0.6	0.0	0.3	0.3	0.0	3.3	0.0

	Me 1	Me 2	Me 3	Me 4	Me 5
Total time with erection	7.5	6.0	14.0	13.1	16.0

*Total time: 120 min.

alone time of the subadult female increased markedly in comparison to both Experiments 3 and 5 (69 minutes versus 16.5 and 13.5 minutes, respectively). This experiment appears to represent a turning point in the social structure of the colony, as well as in the aggressivity of the young male. Whereas in the first three experiments the dominant role was played by the adult female (Melissa), and Nidia only approached and attempted to mount her from a head-on position, during Experiment 4 he managed for the first time to subdue and mount her in the normal fashion, as well as to copulate with her. The earlier close relationship between the two females was interrupted by him, followed by his attacking and chasing away of the subadult female, Candy. As a consequence of this, she spent most of her time sitting alone and/or climbing. By the fifth experiment, on the other hand, Nidia had firmly established his dominance; but being a relatively shy animal by nature, he no longer aggressed against Candy, and permitted her to rejoin the group. A virtual menage a trois developed, in which Nidia interacted sexually with Melissa while Candy groomed the other female.

Some findings utilizing telemetered recording of electrical activity of limbic and cortical regions through permanently implanted electrodes in monkeys are shown in Figures 7 and 8. These animals were injected with methaqualone at the same dosage level as that used in the above-described social interaction experiments. A few minutes after IM injections, high amplitude four to five c/s slow waves appeared in the amygdala and the hippocampus, together with spiking, especially in the amyygdala (Figure 7). The slow activity reached the neocortex later. In the limbic system, the slowing showed its peak actively approximately 30 minutes after injection, whereas relatively minor changes occurred in the neocortical regions. The significance of these findings will be discussed in connection with possible site and mode of action of methaqualone and how this may relate to the results of the behavioral observations.

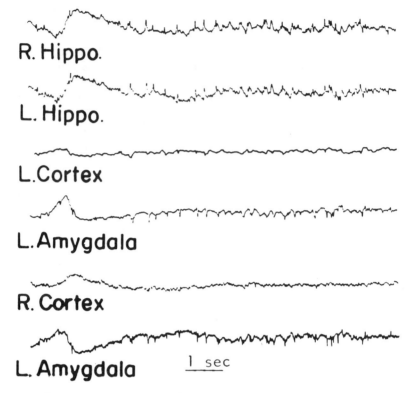

Figure 7. Multi-channeled telemetered recording from freely moving M. mulatta. Spiking in limbic structures begins three min following injection of methaqualone.

DISCUSSION

On initiating the first experimental series, we posed eight questions, delineated earlier, to all but two of which the answers were positive. We did not find any permanent changes in the social rank of the experimental animals or in their behaviors, once the effects of the drug had worn off.

Of the small number of papers in the literature which report experiments utilizing social groups of monkeys for the study of drug effects, only two, both by Redmond and his coworkers, take into account changes in the colony as a whole, resulting from administration of drugs to individual animals (Redmond et al., 1971b). These investigators, following dosing of monkeys with alpha-methyl-paratyrosine -- a known inhibitor of catecholamine

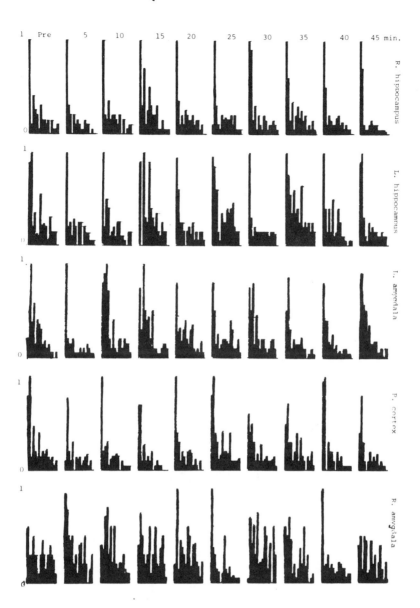

Figure 8. Quantitative EEG analysis from telemetered recording of free moving M. mulatta. First column: before injection of methaqualone, subsequent columns: after five minute intervals. Each abscissa from 2 to 32 Hz. Ordinate: amplitude from 0 to 1. Note that after five minutes post injection, there is spiking in the subcortical structures, which reaches the cortex only at about 30 minutes.

synthesis -- found that quantitative social interactions and appearance did change with treatment. The two chronically drugged animals initiated fewer social interactions, appeared withdrawn, and showed a change in social rank in a downward direction, whereas the altered behaviors were reversible on cessation of treatment, and one of the two experimental animals regained its original rank in the colony order. In a second series, both alpha-methyl-paratyrosine (AMPT), and the indoleamine, p-chlorophenylalanine (PCPA) were tested (Redmond et al., 1971b). Although the animals depleted of serotonin by PCPA treatment eventually developed various physical disabilities, no behavioral alterations were seen. On the other hand, the AMPT treated animals exhibited the same behavioral changes as in the earlier experiments, and two of the four test animals showed a permanent change in social rank in a downward direction. Both of these studies involved chronic administration of monoamine inhibitors, rather then acute treatment with drugs having abuse potential, but the attention paid by the investigators to the effect of the drug on rank within the social order makes them unique in the literature on drug studies in monkey colonies.

The work of Crowley and his co-authors (1974), referred to earlier, represents a major attempt to utilize an intact monkey colony to study the effects of widely abused drugs on the behaviors of individual animals in a social setting. They tested four substances, three of which are CNS depressants (ethanol, pentobarbital, and morphine) and one a psychoanaleptic agent (metamphetamine). They carried out a careful dose-response study of each compound on five male animals in a colony of 30; and they reported drug-specific and dose-dependent behavioral changes in the target animals. However, the doses employed were rather low. Apparently the authors extrapolated mg/kg quantities from man to their animals, not taking into account the much higher metabolic rate of the monkeys. For instance, their highest pentobarbital administration was 1 mg/kg, which would not even approach the usual hypnotic dose in man (100 mg) and certainly is way below the amounts taken by abusers. In comparison, in our studies, the 10 mg/kg of methaqualone employed would correspond to an anesthetic level in man; whereas in the animals, this was the lowest level at which mild ataxia was observed in the initial dose-response studies.

The graphs of Crowley et al., (1974) represent means of behaviors of the five subjects at each drug level. As Hinde pointed out in his monograph (1972): "It is necessary for the student of behavior to focus on individual differences as well as on characters common to groups" (p.12). In a recent study by Miller and Geiger (1976) dealing with the effects of amphetamine on the social behavior on macaques, the authors noted extreme variability in drug response, dependent upon the individual characteristics of the test animal. They utilized 1 and 4 mg/kg of amphetamine -- an order of magnitude higher than the schedule of Crowley et al., (1974). Since they found identifiable drug effects on behavior, which were not uniform from animal to animal, they described individual profiles, instead of calculating means. Our findings also suggested that the behavioral characteristics of an animal had an important influence on the drug effect. This was borne out by our second experimental series (Claus et al., in press).

There seems to be some discrepancy between our first experimental results (Claus et al., 1980) and those reported in our second paper (Claus et al., 1981) concerning the sexual activity of males. In the first publication we described autoerotic behavior (masturbation or autofellatio) in both males, starting after 80 to 100 minutes post drug injection. Although there were eight females in that colony, we did not observe copulation. On the other hand, in the second experimental series, we saw neither active masturbation nor autofellatio on the part of the naive male, but found that he learned how to mount and eventually to copulate, and that he allowed himself to be masturbated by the older female. There are two, possibly simultaneous, explanations for this seeming discrepancy. First, the colony structure described in the first paper was quite different. Big Daddy, the dominant male, showed little interest in any particular female, although he occasionally ceremoniously mounted one of them. Neither during the experiments, not during the baseline or saline studies, did we ever observe him in active copulation. On the other hand, when the younger male was dosed, he actively pursued several of the females but was either chased away or was discouraged by the intervention of Big Daddy. Thus, in his aroused state, he was practically forced to resort to autoeroticism. In the early experiments, only one animal at a time was dosed, whereas in the second series all three subjects were treated simultaneously, and the data suggest that the females were as receptive as the male to close bodily contact and to sexual approaches. For example, in Experiments 1, 2, and 3, Melissa mounted Candy several times and constantly elicited grooming from her. In Experiment 3, Nidia sat with Melissa with erect penis, whereupon she started to masturbate him.

In our first report, we also speculated that the aggressive behavior in the second phase of the experiment shown by the only female receiving methaqualone might be an expression of a sex difference in drug response. Our newer findings seem to rule out this possibility, since marked aggressivity was shown only by the male, and this only until he had established his dominance. Thus, as has been pointed out by Miller and Geiger (1976), it is essential to take into account the individual profiles of the animals in evaluating their responses to psychoactive drugs. During the baseline studies, if one takes note of the fundamental nature of the individual subjects -- whether or not they are loners, or striving to achieve dominance, or are aggressive towards those lower in the ranking order, docile, etc. -- these characteristics will be seen to influence their behaviors under the influence of psychoactive agents.

The effects of methaqualone, as alluded to earlier, are definitely biphasic. Within 10 minutes following injection, the animals became ataxic, and remained stable for the next 70 to 90 minutes, at which point other behavioral changes became manifest, referred to above as the second phase. This biphasic effect of methaqualone on the behavior of the subjects reported earlier has also been borne out by the second experimental series. Whereas during the first 80 to 100 minutes the animals engaged mainly in such affiliative behaviors as sitting together, grooming, etc. True sexual activity and, in some instances aggressive behavior, was seen only in the last 20 to 40 minutes.

Earlier studies conducted to establish time curves for the distribution of methaqualone in rat brain showed that subcortical structures take up the drug faster and in higher quantities than cortical areas (Claus et al., 1978). The highest accumulation after incubation in rats was observed in the optic chiasm and the reticular formation, followed by the midbrain. If uptake in the monkey is similar to that reported in the rat, the ataxia observed shortly after injection probably originates from a disruption of polysynaptic conduction in the reticular formation (Smythies, 1970), rather than from the medulla, as postulated by Swift et al. (1960). Indeed, the studies of Bhargava and his coworkers (1972) showed that the muscle relaxant activity of methaqualone is related to its "greater affinity for polysynaptic pathways than the spinal cord" (p. 806). In the rat experiments (Claus et al., 1978), onset of concentration of the drug was delayed in the cortex relative to that in the reticular formation, while its decline from cortical regions was rapid, especially in comparison to the midbrain. Relatively late high concentrations in the midbrain might account for the altered behaviors observed in the second phase of the present experiments with monkeys.

These suppositions seem to be supported by our findings utilizing telemetered recordings of electrical activity of limbic and cortical regions, through permanently implanted electrodes in monkeys. The animals injected with methaqualone at the same dosage level as that utilized in the social interaction experiments show early high amplitude 4-5 c/s slow waves in the amygdala and hippocampus together with spiking. The slow activity reaches the neocortex later. In comparison, diazepam produces a generalized low amplitude slow wave activity in the amygdala and hippocampus, simultaneous with the appearance of a flattened pattern in the cortex (unpublished). These data seem to indicate that methaqualone has a rather specific action on brain activity, quite unlike that exhibited by diazepam.

The first experimental series suggested the possibility that methaqualone has an aphrodisiac effect. While no copulation was observed among the animals under the influence of the drug, it did act to produce autoerotic behavior in the males. Other similarities to reported human behavior under the influence of the compounds can be gleaned from the studies. The treated male subjects engaged more actively in affiliative behaviors, seeking close body contact with greater frequency under the control conditions.

The second experiments in which three animals were dosed concurrently seem to confirm that methaqualone has an aphrodisiac effect -- manifested especially in the naive male -- which first promotes close body contact and eventually leads to overt sexual activity. Thus, the anecdotal reports from human users of the drug's aphrodisiac qualities seem to be substantiated by these findings with primates.

ACKNOWLEDGEMENTS

This research was supported by NIMH Interdisciplinary Training Grant # MH 15785-01.

NOTES

Abbreviations used in this paper: Ag, aggressive behavior (theatening, slapping, displacing); Ap, approaching; Au, autofellatio; Bl, balancing; Cl, climbing; Dis, displacing; Ea, eating (usually combined with sitting alone); Fa, falling; Fi, fighting; Fr, friendly behavior (approaching, mounting, presenting); Gr, grooming; Hu, huddling; In; inspecting; Ma, masturbating; Mo, mounting; Na, napping: Pa, pacing; Pr, presenting; Sa, sitting alone; Si, slipping; Sl, slapping; Sw, sitting with;

Th, threatening; UCl, unsuccessful climbing; Wa, walking;*by (passive, for example, *Ap, approached by) Ba, baseline; Beh, behavior; Me, methaqualone; S, saline.

REFERENCES

Bhargava, K.P., Rastogi, S.K., and Inha, J.M. The muscle relaxant activity of methaqualone and its methylcongener. Br. J. Pharmacol. 44, 805-806 (1972).

Claus, G., DeBernardo, E., and Krisko, I. Pilot study on the distribution of ^{14}C-labeled methaqualone in the rat brain. Biochem. Pharmac. 27, 1300-1303 (1978).

Claus, G., Kling, A., and Bolander, K. Effects of methaqualone on social behavior in monkeys (M. mulatta). Brain Beh. Evol. 17, 391-410 (1980).

Claus, G., Kling, A., and Bolander, K. Effects of methaqualone on social-sexual behavior in monkeys (M. mulatta), part II. Brain Beh. Evol. 18, 105-113 (1981).

Cressman, R.J. and Cadell, T.E. Drinking and the social behavior of rhesus monkeys. Q. J. Stud. Alcohol. 32, 764-774 (1971).

Crowley, T.J., Stynes, A.J., Hydinger, M., and Kaufman, C. Ethanol, methamphetamine, pentobarbital, morphine and monkey social behavior. Arch. Gen. Psychiatry 31, 829-838 (1974).

Derbez, R. and Grauer, H. A sleep study and investigation of a new hypnotic compound in a geriatric population. Can. Med. Ass. J. 97, 1389-1393 (1967).

Deutsch, S. Fifteen-electrodetime-multiplex EEG telemetry from ambulatory patients. IEEE Trans. Biomed. Eng. BME-26, 153-159 (1979).

Gamage, J.R. and Zerkine, E.L. Methaqualone. Natl. Clearinghouse Drug Abuse Info. 18, 1-13 (1973).

Gelpke, R. Drogen und Seelenerweiterung. Kindler Verlag, Munich (1975).

Gerald, M.C. and Schwirian, P.M. Nonmedical use of methaqualone. Arch. Gen. Psychiatry 28, 627-631 (1973).

Hinde, R.A. Social Behavior and its Development in Subhuman Primates. Oregon State System of Higher Education, Eugene, Ore. (1972).
Inaba, D.S., Gay, G.G., Newmeyer, J.A., and Whitehead, C. Methaqualone abuse: "Luding out". J. Am. Med. Ass. 224, 1505-1509 (1973).
Kackher, J.K. and Zaheer, S.H. Potential analgesics. I. Synthesis of substituted 4-quinazolones. J. Indian Chem. Soc. 28, 344-346 (1951).
Kaufman, I.C. and Rosenblum, L. A behavioral taxonomy for Macaca nemestrina and Macaca radiata. Primates 7, 205-258 (1966).
Kjellberg, B. and Randrup, A. The effect of amphetamine and pimozide, a neuroleptic, on the social behaviour of vervet monkeys (Cercopithecus sp.), in Advances in Neuro-Psychopharmacology, O. Vinar, Z. Votava, and P.B. Bradley, eds. North Holland, Amsterdam (1971), pp. 305-310.
Kling, A., Steklis, D., and Deutsch, S. Radiotelemetered activity from the amygdala during social interactions in the monkey. Exper. Neurol. 66, 88-96 (1979).
Kohli, R.P., Sing, N., and Kulshrestha, V.K. An experimental investigation of dependence liability of methaqualone in rats. Psychopharmacologia 35, 327-334 (1974).
Lewis, J.R. and Steindler, E.M. Methaqualone. J. Am. Med. Ass. 224, 1521-1522 (1973).
MacAndrew, C. and Edgerton, R.B. Drunken Comportment. Aldine Press, Chicago (1969).
Miller, M.H. and Geiger, R. Dose effects of amphetamine on macaque social behavior: Reversal by haloperidol. Res. Comm. Psychol. Psychiatry Beh. 1, 125-142 (1976).
Miller, R.E., Levine, J.M., and Mirsky, A.I. Effects of psychoactive drugs on non-verbal communication and group social behavior of monkeys. J. Pers. Soc. Psychol. 28, 396-405 (1973).
Redmond, D.E., Maas, J.W., Kling, A., and Dekirmenjian, H. Changes in primate social behavior after treatment with alpha-methylparatyrosine. Psychosomatic Med. 33, 97-113 (1971a).

Redmond, D.E., Maas, J.W., Kling, A., Graham, C.W., and Dekirmenjian, H. Social behavior of monkeys selectively depleted of monoamines. Science 174, 428-431 (1971b).

Sargant, W. Prescribing Mandrax. Brit. Med. J. 2, 716 (1973).

Saxena, R.C., Bhatnagar, N.S., Misra, S.C., and Bhargava, K.P. Intravenous methaqualone: A new non-barbiturate anaesthetic. Brit. J. Anaesth. 44, 83-85 (1972).

Smythies, J.R. Brain Mechanism and Behavior. Blackwell Sci. Publ., Oxford (1970).

Soulairac, A. and Gottesmann, C. Experimental studies on sleep produced by methaqualone. Life Sci. 6, 1229-1232 (1967).

Swift, J.B., Dickens, E.A., and Becker, B.A. Anticonvulsant and other pharmacological activities of Tuazolone. Arch, Gen. Int. Pharmacodyn. Ther. 128, 112-125 (1960).

Chapter 9
Influence of Amphetamine and Neuroleptics on the Social Behavior of Vervet Monkeys

I. Munkvad and A. Randrup

About 1966 our laboratory put forward the hypothesis that hyperactivity of the dopaminergic system in corpus striatum played a role in the pathogenesis of schizophrenia (Randrup and Munkvad, 1963, 1965a, 1965b, 1966, 1967b, 1967c, 1968, 1972; Munkvad et al., 1968, 1970; Fog et al., 1966; Fog, 1972). The hypothesis, now called the dopamine-hypothesis of schizophrenia, was based on the fact that in the clinic amphetamine and related compounds may provoke a schizophrenia-like psychosis with all schizophrenic symptoms persisting for a rather long time -- several months up to half a year, even when the intake of amphetamine was stopped on admission to the hospital.

In animal studies amphetamine stereotypies have been described in many species, including vervet monkeys, where this abnormal behavior is more individual with several variations from animal to animal.

Social behavior under the influence of amphetamine and/or neuroleptics has been studied with two or three monkeys together in a cage, and these studies will be reviewed here.

All known psychotic drugs currently in use have an anti-dopaminergic action, and in animal experiments they are able to abolish or diminish amphetamine-induced stereotyped activity.

From many animal experiments it has been stated that an intact corpus striatum and the presence of the neurotransmitter dopamine is necessary for developing amphetamine-induced stereotyped behavior. In our experiments stereotypy is defined as pathological behavior produced by psychoactive drugs such as amphetamines, with some persisting meaningless repetitive behavior, while normal behavior is abolished or diminished.

As already mentioned, amphetamine has a strong influence on behavior in animals, including social behavior. In vervet monkeys we have found significant effects on social behavior with doses of amphetamine as low as 0.05 mg/kg body weight. Three different social situations were studied:
  (1) Two monkeys, male and female
  (2) Three monkeys, one male and two females
  (3) Mother and infant

The first experiments (two monkeys, male and female) were performed in the laboratories of Ferrosan Ltd., Malmö, Sweden, in cooperation with B. Kjellberg (Kjellberg and Randrup, 1971, 1972a, 1973; Randrup et al., 1976, 1980). Social behavior such as mutual grooming, sexual intercourse, and other forms of touching and vocalization were recorded in six pairs of monkeys. Changes in these behaviors, most often reductions, were seen after injections of d-amphetamine to both monkeys. Significant changes were found after a dose of 0.05 mg/kg body weight as mentioned above (Table 1), with increasing effects at higher doses, such as 0.15 and 0.37 mg/kg body weight.

Other experiments (three monkeys, one male and two females) were performed in our laboratories at St. Hans Mental Hospital in cooperation with E. Schiørring (Schiørring, 1972, 1977, 1979). Amphetamine in doses of 0.1-0.7 mg/kg body weight (given to each of the three monkeys) changed both the individual and the social behavior patterns significantly. Stereotypy and social isolation (withdrawal) were characteristic features of the amphetaminized animals. Stereotyped social grooming was also observed. Three independent groups were investigated.

Figure 1 shows results from the first group. A line drawn between two symbols shows that social grooming occurred between the two monkeys denoted by the symbols. The triangle at the bottom shows that all the possible grooming relations occurred in each of the 16 placebo experiments, indicating a stable pattern of social interaction. This pattern was clearly disturbed in 12 of the amphetamine

Table 1. Effects of 0.05 mg/kg s.c. d-amphetamine on social behaviour of pairs of vervet monkeys

	A + B Placebo	A + B Amphetamine	H + O Placebo	H + O Amphetamine	R + Ju Placebo	R + Ju Amphetamine
♂grooming♀	14.3	1.5*	2.3	0**	3.7	2.2
♀grooming♂	6.0	5.3	4.7	0**	6.1	2.7
♂stretching at♀	5.8	10.5*	4.8	0.2**	7.6	4.0
♀stretching at♂	6.6	2.0	0.7	0*	1.7	0.7
♀presents rear quarters to♂	3.4	0**	0.6	0	3.7	1.7
Two-tone sound by ♀	0.8	6.8				
No. of experimental days	12	6	12	6	12	6

The figures in the table represent average number of times the various behavioural items were performed per hour. Very similar results were obtained from calculations based on the duration of the behavioural items (in seconds). There were two 30 min observation periods each experimental day, starting 30 and 75 min after the injection. Placebo experiments were always made the day before and the day after amphetamine. There were 2-5 weeks between amphetamine injections so all observations on each pair were made within 4 months. One previous experiment with A + B was confirmatory (Kjellberg and Randrup, 1973).

*Significantly different from placebo, $P < 0.05$.   **Significantly different from placebo, $P < 0.005$

240 MUNKVAD & RANDRUP

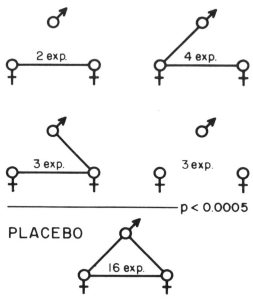

Figure 1. (Group 1) Mutual grooming.

experiments as shown in the upper part of the figure. In the remaining four amphetamine experiments all the grooming relations occurred but the frequencies were different from placebo. In the amphetamine sessions, the social grooming was replaced by stereotypies or staring into space. Level of significance (P<0.0005) was calculated by the Fisher exact test based on presence or absence of all three grooming relations between two monkeys. In each of the 16 placebo experiments there were also episodes where all three monkeys were involved simultaneously in mutual grooming (average duration 659 s.). Such episodes were seen only in 7 of the 16 amphetamine experiments (average duration 33 s.). Calculated (Fisher exact test) as above, this difference is significant (P<0.003).

In Group II one of the grooming relations (a male grooming one of the females) was strongly increased by amphetamine. The development of such social stereotypies has also been observed in a few other monkey experiments and in abusers of amphetamine and cocaine.

In the last group the most prominent effect of amphetamine was total isolation. All three monkeys were sitting immobile (but with head movements) at separate places in the cage, staring continuously at the observation window.

In the third type of experiment (mother and infant) doses of d-amphetamine of 0.1, 0.15, and 0.2 mg/kg body weight were used and given to the mother only (Schiørring, 1977; Schiørring and Hecht, 1979).

The fundamental parental care behavior pattern was disrupted, and the mother became isolated in a social withdrawn phase. She did not respond to the calling signals of the infant and showed behavior in which stereotyped self-grooming and/or staring into space were predominant.

The reactions of the infant to the amphetamine-induced behavior of the mother were different in the two experimental pairs investigated. In Group I the infant increased its approach-avoidance movements. In Group II, the infant sat very quietly and close, in front of the mother.

Table 2 shows the results from a special part of the experiment, where mother and infant were temporarily separated by a clear fiberglass barrier.

The mother from Group I reacted to the increased approach from its infant with active rejection. In both groups the mother failed to react with the typical ventral-ventral grasping either with the infants sitting close or to the social anxiety signals of the infants. In all placebo sessions the mother reacted adequately to the calling signals of the infant and spent 91.7% of the total observation time close (less than 10 cm) to the separating wall. After the nonchronic amphetamine treatment the mother spent 91.4% of total observation time away from the wall (more than 2 m) preoccupied with stereotyped self-grooming, looking at her hands, picking on the left hand continuously, and biting nails. The significance of her social

GROUP I

MOTHER/INFANT
SOCIAL INTERACTION VERVET MONKEYS

% OF TOTAL SESSION TIME (1 HR.)

	MOTHER AWAY FROM SCREEN	MOTHER CLOSE TO SCREEN	INFANT CLOSE TO SCREEN
PLACEBO N=15	8.3	91.7	78.7
0.2 MG KG AMPHETAMINE TO MOTHER N=15	91.4	8.6	75.0

P < 0.0005

STEREOTYPES: LOOKING AT HANDS
PICKING ON LEFT HAND
BITING NAILS

SOCIAL WITHDRAWAL: NO RESPONSE TO INFANTS CALLING-SIGNALS

Table 2. Response of the Mother to the Infant in the Social Cohesion Test

withdrawal was that she stopped reacting to the infant's calling signals. In spite of differences in behavior changes induced by amphetamine, the main conclusion is that the mothers totally lost their normal and biologically significant interest in their infants.

Neuroleptics given to normal animals reduce social behaviors. This may be one of the reasons why full reversal of the amphetamine effects has not been obtained (Randrup et al., 1976). Table 3 shows the suppression of social behavior of pairs of monkeys (male and female) by amphetamine and partial prevention by neuroleptics. Each experiment comprised three or four (always same number within one horizontal line) 30 minute observation periods distributed between ½ and 3½ hours after the s.c. injection of 0.37 mg amphetamine (A) per kg. The neuroleptics (N), haloperidol and chlorpromazine,

Monkey pairs	Neuroleptic	Dose[a] (mg/kg s.c.)	No. of diff. items of social behaviour			No. of experiments		
			P	A	N + A	P	A	N + A
A+B	Pimozide	0.03 - 0.06	9.5	0	0.7	6	3	3
J+Si	Pimozide	0.03 - 0.06	6.7	0.3	4.0	6	3	3
O+So	Pimozide	0.03 - 0.06	4.9	1.7	2.2	9	3	6
Jo+Ju	Haloper.	0.015-0.04	5.8	0.3	2.0	13	4	9
G+V	Haloper.	0.015-0.04	6.1	0.7	1.3	9	3	6
A+B	Haloper.	0.015-0.04	6.4	1.0	0.7	5	2	3
Au+Aua	Haloper.	0.015-0.04	6.0	0	0.7	7	4	3
O+So	Chlorprom.	0.2 -0.6	6.4	1.5	4.6	12	4	8
A+B	Chlorprom.	0.2 -0.6	7.5	0	0.3	8	2	6
G+V	Chlorprom.	0.2 -0.6[b]	7.2	0	2.3	9	2	7

[a]Doses were regulated so that amphetamine-stereotypies were just inhibited and no "extrapyramidal" effects of the neuroleptics appeared; amphetamine gives a surmountable protection against the latter.
[b]Monkey V was given up to 1.5 mg chlorpromazine/kg

Table 3. Suppression of Social Behaviour of Pairs of Monkeys (male and female) by Amphetamine and Partial Prevention by Neuroleptics

were injected ½ hour, and pimozide 3 hours before amphetamine. A placebo (P) experiment was performed on the day before each drug experiment. The social behaviors comprise grooming of the other monkey, stretching to be groomed, touching the other monkey with the hand, presenting rear to male, mounting, fighting, and biting. Social behaviors were counted in each experiment and the averages are shown in Table 3. Total occurrences of social behaviors and total time spent in these behaviors were also evaluated and show the same trend. Mutual grooming was the predominant behavior (Kjellberg and Randrup, 1971, 1972b, 1973; Randrup et al., 1976; Munkvad et al., 1980).

Accumulating evidence from other quarters supports the above-mentioned results both with amphetamine, neuroleptics, and the combination (Crowley et al., 1974; Garver et al., 1975;

Machiyama et al., 1970, 1975; Miller, 1976; Haber et al., 1977; Ridley et al., 1979; Scraggs et al., 1979; Baker and Ridley, 1979).

In our opinion the experiments reported here show strong parallels to some behavior of schizophrenic patients. Autism or social withdrawal is one of the main traits in the schizophrenic symptomatology, and the social isolation of vervet monkeys under the influence of amphetamine has many common traits with the autism seen in schizoprenic patients, even though it can be misleading to draw direct parallels from animal experiments to the human clinic. Nevertheless, we feel that the experiments described in relation to social behavior of vervet monkeys represent an animal model of autism and can be used for further investigation in relation to the pathogenesis of schizophrenia.

The finding that neuroleptics attenuate positive social behavior in normal nonhuman primates has implications for the clinical use of these drugs. Behavioral and psychotherapeutic treatment concurrent with neuroleptics may impair these therapies by virtue of the neuroleptics inhibiting the replacement of negative behaviors by positive ones in schizophrenic patients.

REFERENCES

Baker, H.F. and Ridley, R.M. Behavioural effects of chronic amphetamine and their reversal by haloperidol in the marmoset. Brit. J. Pharmacol. 66, 146P-147P (1979).

Crowley, T.J., Stynes, A.J., Hydinger, M., and Kaufman, I.C. Ethanol, methamphetamine, pentobarbital, morphine, and monkey social behavior. Arch. Gen. Psychiatry 31, 829-838 (1974).

Fog, R.L., Randrup, A., and Pakkenberg, H. Amines in the corpus striatum associated with the effects of both amphetamine and antipsychotic drugs, in Proceedings of the IVth World Contress of Psychiatry, part 2, J.J. Lopez Ibor, ed. Excerpta Medica Internat. Congres. Ser. No. 150, Amsterdam (1966), pp. 2580-2582.

Fog, R. On stereotypy and catalepsy: studies on the effect of amphetamines and neuroleptics in rats. *Acta Neurol. Scandinav.*, suppl. 50, vol. 48 (Thesis) (1972).

Garver, D.L., Schlemmer, R.V., Maas, J.W., and Davis, J.M. A schizophreniform behavioral psychosis mediated by dopamine. *Am. J. Psychiatry* 132, 33-38 (1975).

Haber, S., Barchas, P.R., and Barchas, J.D. Effects of amphetamine on social behaviors of rhesus macaques: An animal model of paranoia, in *Animals Models in Psychiatry and Neurology*, I. Hanin and E. Usdin, eds. Pergamon Press, Oxford (1977), pp. 107-115.

Kjellberg, B. and Randrup, A. The effect of amphetamine and pimozide, a neuroleptic, on the social behaviour of vervet monkeys (Cercopithecus sp.), in *Advances in Neuro-Psychopharmacology*, O. Vinar, Z. Votava, and P.B. Bradley, eds. North Holland, Amsterdam (1971), pp. 305-310.

Kjellberg, B. and Randrup, A. Changes in social behaviour in pairs of vervet monkeys (Cercopithecus) produced by single, low doses of amphetamine. *Psychopharmacologia* 26, 117, suppl. (1972).

Kjellberg, B. and Randrup, A. Stereotypy with selective stimulation of certain items of behaviour observed in amphetamine treated monkeys (Cercopithecus). *Pharmakopsychiatrie Neuro-Psychopharmakol.* 5, 1-12 (1972).

Kjellberg, B. and Randrup, A. Disruption of social behaviour of vervet monkeys (Cercopithecus) by low doses of amphetamines. *Pharmakopsychiat.* 6, 287-293 (1973).

Machiyama, Y., Utena, H., and Kikuchi, M. Behavioural disorders in Japanese monkeys produced by the long-term administration of methamphetamine. *Proc. Jap. Acad.* 46, 738- (1970).

Machiyama, Y., Hsu, S.C., Utena, H., Katagiri, M., and Kurata, A. Aberrant social behaviour induced in monkeys by the chronic methamphetamine administration as a model for schizophrenia, in *Schizophrenia and Schizophrenia-Like Psychoses*, H. Mutsuda and T. Fukuda, eds. Georg Thieme Publ., Stuttgart, and Igaku Shoin Ltd., Tokyo (1975), pp. 97-108.

Miller, M.H. Behavioural effects of amphetamine in a group of rhesus monkeys with lesions of dorsolateral frontal cortex. Psychopharmacology 47, 71-74 (1976).

Munkvad, I. and Randrup, A. Evidence indicating the role of brain dopamine in the psychopharmacology of schizophrenic psychoses. Psihofarmakologija 2, 45-47 (1970).

Munkvad, I., Pakkenberg, H., and Randrup, A. Aminergic systems in basal ganglia associated with stereotyped hyperactive behaviour and catalepsy. Brain, Behav., Evol. 1, 89-100 (1968).

Munkvad, I., Randrup, A., and Fog, R. Amphetamines and psychosis, in Hormones and the Brain, D. de Wied and P.A. van Keep, eds. MTP Press, England (1980), pp. 221-229.

Randrup, A. and Munkvad, I. On the relation of adrenergic and tryptaminic mechanisms to amphetamine-induced abnormal behaviour. Scand. J. Clin. Lab. Invest. 14, 44 (1963).

Randrup, A. and Munkvad, I. Pharmacological and biochemical investigations of amphetamine-induced abnormal behaviour, in Neuro-Psychopharmacology 4, D. Bente and P.B. Bradley, eds. Elsevier, Amsterdam (1965a), pp. 301-304.

Randrup, A. and Munkvad, I. Special antagonism of amphetamine-induced abnormal behaviour. Psychopharmacologia 7, 416-422 (1965b).

Randrup, A. and Munkvad, I. Role of catecholamines in the amphetamine excitatory response. Nature 211, 540 (1966).

Randrup, A. and Munkvad, I. Bran dopamine and amphetamine-induced stereotyped behaviour. Acta Pharmacol. Toxicol. 25, suppl 4, 62 (1967a).

Randrup, A. and Munkvad, I. Stereotyped activities produced by amphetamine in several animal species and in man. Psychopharmacologia 11, 300-310 (1967b).

Randrup, A. and Munkvad, I. Stereotyped behavior produced by amphetamine and other substances, in Neuro-Psychopharmacology 5, H. Brill, J.O. Cole, P. Deniker, H. Hippius, and P.B. Bradley, eds. Excerpta Medica Internat. Congr. Ser. No. 129, Amsterdam (1967c), p. 1225.

Randrup, A. and Munkvad, I. Behavioural stereotypies induced by pharmacological agents. Pharmakopsychiatrie Neuro-Psychopharmakol. 1, 18-26 (1968).

Randrup, A. and Munkvad, I. Evidence indicating an association between schizophrenia and dopaminergic hyperactivity in the brain. Orthomolecular Psychiatry 1, 2-7 (1972).

Randrup, A., Munkvad, I., Fog, R., Kjellberg, B., Lyon, M., Nielsen, E., Svennild, I., and Schiørring, E. Behavioural correlates to antipsychotic efficacy of neuroleptic drugs, in International Symposium on "Antipsychotic Drugs, Pharmacodynamics and Pharmacokinetics", G. Sedval, B. Uvnäs, and Y. Zotterman, eds. Pergamon Press, Oxford (1976), pp. 33-41.

Randrup, A., Kjellberg, B., Schiørring, E., Scheel-Krüger, J., Fog, R., and Munkvad, I. Stereotypies and their relevance for testing neuroleptics, in Handbook of Experimental Pharmacology, Vol. 55: Psychotropic Agents-Antipsychotics and Antidepressants, F. Hoffmeister and G. Stille, eds. Springer Verlag, Berlin (1980), pp. 97-110.

Ridley, R.M., Scraggs, P.R., and Baker, H.F. Modification of the behavioural effects of amphetamine by a GABA agonist in a primate species. Psychopharmacology 64, 197-200 (1979).

Schiørring, E. Social isolation and other behavioural changes in a group of three vervet monkeys (Cercopithecus) produced by single, low doses of amphetamine. Psychopharmacologia 26, 117, suppl. (1972).

Schiørring, E. Changes in individual and social behavior induced by amphetamine and related compounds in monkeys and man, in Advances in Behavioral Biology 21, E.H. Ellinwood and M.M. Kilbey, eds. Plenum Press, New York (1977), pp. 481-522.

Schiørring, E. Social isolation and other behavioural changes in groups of three adult vervet monkeys (Cercopithecus aethiops) produced by low, non-chronical doses of d-amphetamine. Psychopharmacology 64, 297-302 (1979).

Schiørring, E. and Hecht, A. Behavioural effects of low, acute doses of d-amphetamine on the dyadic interaction beteen mother and infant vervet monkeys (Cercopithecus aethiops) during the first six postnatal months. Psychopharmacology 64, 219-224 (1979).

Scraggs, P.R., Baker, H.F., and Ridley, R.J. Interaction of apomorphine and haloperidol: Effects on locomotion and other behaviour in the marmoset. Psychopharmacology 66, 41-43 (1979).

Chapter 10
Effects of Drugs on the Response to Social Separation in Rhesus Monkeys

W.T. McKinney, Jr., E.C. Moran, and G.W. Kraemer

INTRODUCTION

One of the proposed animal models for the study of human depression involves experimental separation of nonhuman primates from their main affectional objects. This chapter summarizes the behavioral aspects of these peer separation models, and briefly describes the results of recent studies of various pharmacological agents which interact with this response to separation.

Several methods have been used to create animal models of depression. They are based on different conceptual schemes and involve varied preparations, including both drug and socially induced syndromes in primate and rodent species. At the present time, no single animal model perfectly mirrors human depression, but there are several reasonable ones available. Each of these models has its own advantages and limitations. However, since the specific intent of this chapter is to discuss separation models and the effects of a number of agents on the response to separation, discussion of other kinds of models will be done only as they relate to the peer separation model.

NONHUMAN PRIMATE SOCIAL SEPARATION MODELS

"Despair" is a term used to describe the severe syndrome which often occurs following separation. Rhesus monkeys in the despair stage may evince symptoms of life-threatening illness including failure to maintain adequate food and water intake, lethargy, withdrawal, and insensitivity to external stimuli. Death can occur even when the investigator intervenes to maintain adequate fluid and electrolyte balance. Autopsy typically reveals nothing to explain the death. There are basically two types of social separation models currently being used in experimental studies of depression. One of these is mother-infant separation and the other is peer-peer separation.

MOTHER-INFANT SEPARATION

Spitz (1946), Bowlby (1960), and the Robertsons (1952, 1955, 1971) observed children in institutions (usually hospitals or nurseries) where they were separated from their mothers and families. They originally described human infants' reactions to separations and used the terms "protest" and "despair" to describe the resulting behaviors. These terms were later used by others to describe infant monkeys' response to separation from their mothers.

The first studies of experimental separation of monkey infants from their mothers were reported in the 1960's. Seay and Harlow (1965) separated rhesus macaque (Macaca mulatta) infants from their mothers at approximately seven months of age for three weeks. Jensen and Tolman (1962) separated pigtail macaques (M. nemestrina) infants from their mothers at approximately the same age for periods of less than an hour and then returned them to their own or another mother. In both reports, these separations were described as highly stressful events for the infants as evidenced by increased arousal and increased distress vocalizations. Seay and Harlow (1965) compared the infants' reactions to the behavior patterns reported for human children who had been separated from their parents and/or families. They classified the behaviors observed in these infant monkeys into the two categories of "protest" and "despair." The behavior profile for infant monkeys in the protest stage consists of a generalized disoriented increase in activity accompanied by loud and repeated screeching. The despair stage which follows consists of decreased activity, decreased vocalizations, decreased food and water consumption, and general social withdrawal.

As this original work on the response to maternal separation was extended, additional research revealed that the response of the infants was most likely influenced by a number of different factors. When bonnet (M. radiata) infants were separated from their mothers (Rosenblum and Kaufman, 1968), they exhibited only a mild protest reaction and almost no despair response. This is a much different response than the more severe biphasic reaction found in both rhesus and pigtail macaque infants. They speculated that the nature of the social structure of a particular species is the underlying reason for the different response. Bonnet infants spend a much larger percentage of their time interacting with troop members other than their mothers and the general behavior of troop members is more permissive and responsive to the infants than with rhesus or pigtail macaques where the bonds are more dyadic in nature. A severe despair response much more like that of a rhesus or pigtail macaque can be produced even in the bonnet macaque if the infant is deprived of the socializing of the troop members during the time it is separated from its mother (Kaufman and Stynes, 1978).

When Suomi and Harlow (1977) separated rhesus infants from their mothers at 60, 90, and 120 days of age, they found that all subjects showed a typical protest and despair response. However, the infants which had been separated at 90 days reacted more dramatically than those that had been separated at other ages. They also found that infants which were housed in single cage units during the separation period showed a much stronger despair response than infants which were housed with a peer. These differences persisted and even worsened in some cases until six months of age. These data indicate that age and social condition during the time of separation can also have an effect on the despair response.

Sex and preexisting behavior patterns have also been suggested as affecting the response to maternal separation (Hinde and Spencer-Booth, 1970,

1971; Hinde and Davies, 1972; Hinde and McGinnis, 1977). Hinde and colleagues separated rhesus infants from their mothers for short periods of time. They reported the occurrence of the protest and despair stages and presented data which indicated that preseparation behavior can predict behavior which follows separation. They also suggested that sex of the infants may be another variable which influences the response to maternal separation, with male infants showing a more severe depression than female infants.

Most of the work described so far has focused on the behavioral response to maternal separation. Other research has been done to measure concomitant biological parameters. The results of these studies have also tended to support mother-infant separation as a model for some aspects of human depression.

Reite and Short (1978) used remote telemetry recording techniques to record sleep data. The changes they found in the sleep patterns of pigtail infants undergoing separation were very similar to those found in depressed humans.

Other studies (Reite et al., 1974, 1978a, 1978b) found that heart rate and body temperature were elevated in infants that were in the protest stage shortly after being separated from their mothers. By the first evening after separation heart rate, REM sleep, and body temperatures had decreased and remained depressed throughout the despair stages.

Separation also has effects on a variety of neurochemical systems. For example, serotonin levels are elevated in the hypothalamus in rhesus infants that are in the protest stage after separation from their mothers. In this same study, the major enzymes involved in catecholamine synthesis were found to be elevated in the adrenal gland (Breese et al., 1973).

Plasma cortisol levels have been studied as they relate to maternal separation in squirrel monkeys (Saimiri sciureus). This research describes separation, even as short as ½ hour, as being a highly stressful event that is reflected in elevated plasma cortisol levels. Not only were plasma levels elevated, but they stayed elevated regardless of whether the infants were housed alone or with other infants and regardless of whether they were separated from natural mothers or surrogate mothers (Levine et al., 1978; Coe et al., 1978; Smotherman et al., 1979; Vogt and Levine, 1980).

Most of these studies of separation of mother and infant were done to learn more about the nature of the mother-infant dyad itself. However, as this research progressed, it became apparent that a very strong and pervasive response occurred on the part of the infants after being separated from their mothers, and that this response was, in many ways, comparable to those forms of human depression which occur after separation from an affectional object. This observation prompted many researchers to study the maternal separation response itself as a potential animal model of depression.

RATIONALE

Before continuing with more discussion and review of data on the separation model as produced with monkey peers, perhaps a brief discussion of the connections between separation and depression in humans is in order.

There are many theories regarding the role of separation in human depression. Some feel that it is a major causative event while others feel that the relationship is a more complex one with the role of separation being a contributing factor along with other mechanisms, both physiological and social, rather than being a single precipitating event. Although the relationship between separation and depression is complex and intricate, it is generally thought to be a close one. Clinical theories have long postulated that separation or object loss is an important factor in development of depression in humans. More recent empirical studies in this area have been the subject of a recent review (Lloyd, 1980).

A more current conceptualization is that separation or object loss almost invariably leads to a grief reaction but to severe depression only in vulnerable individuals. Attempts to define this vulnerability in humans are underway and animal models can help greatly in this regard.

Separation models in primates by no means assume that all depressions in humans are caused or precipitated by separations or that separation invariably leads to the development of depression. However, there is considerable evidence, which strongly suggests that disruption of peer-attachment bonds in rhesus monkeys has many similarities to some forms of human depression. These similarities include behavioral likenesses, as well as neuropharmacological and drug response factors.

PEER SEPARATION MODEL

Rhesus monkeys are similar to humans in that they develop strong and complex social bonds. Their societies are held together throughout the life time of the animals by a comparatively sophisticated social system. While the ties of the rhesus mother and infant are very intense, in some ways the longer lasting and complex ties of the peer relationship may be more closely related to the diverse factors involved in human social structure. In the mother-infant relationship, the bond is centered on life support systems for the infant, i.e., nursing, body contact, and protection from outside forces. However, as vital and important as these functions may be, they are, by their very nature, short-lived. Once weaning begins, this role becomes less and less important and as an animal reaches adulthood, its ties with its parent become more and more fragile. The peer relationship, which begins about weaning time and even earlier, continues for the rest of the animal's life. During this time the infant establishes relationships with agemates and other members of a troop, if available, that include play behavior, sex behavior, and food gathering. It is also during this time that dominance hierarchies are formed and that the maturing infant learns to take its place in a group structure.

Peer separation has been proposed by Suomi et al. (1970) and Bowden and McKinney (1972) as an animal model of depression that has many desirable characteristics. Suomi et al. (1970) found that infants who were separated from their mothers, but allowed to remain with their peers, showed similar protest and despair behaviors in response to both the maternal and peer separations. In addition Singh (1975) found protest and despair behaviors in feral infants who had been separated from their mothers and removed from the troop. Their responses were not as striking as that of infants raised in a laboratory, but definite decreases in locomotion, vocalization, environment exploration, and appetite occurred.

The underlying assumptions of the peer separation paradigm as a model of human depression are, in many ways, based on the same phenomenon as that which occurs in the mother-infant separation model. In the latter case, the protest and despair pattern seen as a result of separation from mother is likened to that described for human infants who have been separated from their parents. However, it is equally possible that this behavior as it occurs in peers (especially since it seems to occur in some form in all ages of peers) is equally analogous to that form of human depression which is thought to result from, or be involved with, separation from important affectional objects. More research will help to establish the validity of this assumption. There are characteristics of the peer separation model which, once established, make it a useful and pragmatic tool for some forms of research.

The establishment of a different animal model of depression in peer groups does not disregard the mother-infant model. It is obviously a model of consistency and import and deserves continued study in its own right as well as for a model of depression. However, different aspects of research require different tools, and development of both models may ultimately result in a more complete understanding of monkey and human depression.

There are some pragmatic reasons for using the peer separation model for certain aspects of depression research. The stabler attachment systems are much less subject to change and, therefore, age at the time of separation becomes less important as a variable. In contrast, the rapidly changing nature of the mother-infant relationship usually means that only one separation can be done. This increases the cost of generating data and the animal population needed for a given experiment. Repeated replications of the syndrome are also more easily done in the peer separation model.

The first reported study of repetitive peer separation in monkeys (Suomi, 1970) described a response that was similar to the protest and despair response that has been documented for mother-infant separation. These rhesus infants were separated from their mothers at birth and then housed with peers from 15 days of age. They underwent 20 separations of four days each starting at 90 to 300 days of age. It was found that the infants showed a strong protest and despair response, and that this response was persistent and unabated for 20 repeated separations. The results of this study indicate that frequency of separations is not the decisive factor in influencing the nature of the despair response to peer separation.

The nature of the environmental situation during separation can affect the response to separation (Suomi, 1973). Two groups of rhesus infants were housed during periods of separation either in single cages or in vertical chambers that allowed no visual or physical contact with the immediate surroundings. The infants that spent their time after separation in the vertical chamber were more severely depressed than the infants who spent their time in single cages.

Singh (1977), after separating feral monkeys by removing the rest of the troop from the area and allowing the subject monkey to remain in its home environment reported that the depressive response was not as severe as that seen in the laboratory. He suggested that this was because the subjects needed to watch out for predators. This interpretation supports the contention that characteristics of the environment can affect the severity of the response to separation.

There is some reason to believe that the nature of the infants' preseparation social environment may also influence the infants' later response to subsequent separations. Infants reared with their mothers for three months and then put into peer groups reacted less adversely to separation than infants that had been peer-reared from birth or shortly thereafter. Kraemer (1978) reported that peer-reared subjects showed a stronger response to separation from their peers than mother-reared animals. The mother-reared infants, unlike peer-reared infants, did not have such behaviors as abnormal self-clasping and self-mouthing in their repertoires before separation nor did these behaviors appear after separation. However, other signs of depression did appear; activity and play levels as well as appetite decreased. Mother-reared infants may not show the more bizarre signs of disturbance simply because these behaviors were not part of their original repertoire, or it may be that they simply had not formed as strong attachment bonds to their peers as had the monkeys who had been raised with each other from birth.

Bowden and McKinney (1972) separated, for a two-week period, pairs of juvenile monkeys that had been housed with each other for eight months. They reported signs of mild protest but no despair. The protest was characterized by increased levels of activity and an increase in self-directed behaviors which continued to increase throughout the separation period. There were changes in food and water consumption that began at separation. However, the changes were very erratic with some monkeys showing marked increases while other monkeys showed marked decreases in consumption. A study by McKinney, Suomi, and Harlow (1972) reported similar results when juvenile monkeys (three years old) who had been reared together since infancy but with their mothers present until they were eight months old, were repeatedly separated for four two-week periods. These animals showed increased levels of activity and environmental exploration, i.e., protest-type behaviors, which lasted throughout the separation. There were no reported changes in food and water consumption.

These studies indicate that older animals do react adversely to separation, although they may not show the same patterns as infants that are separated from peers. There appears to be much more variety in the form, duration, and intensity of the response in juvenile monkeys than in infants.

Considered in their entirety, these studies suggest a number of different explanations. The data could be interpreted as indicating that adolescent monkeys do not respond to separation from their peers as intensely as infant monkeys do. It could also be said that these animals were equally affected but that they displayed their disturbance in other ways. Another possibility that must not be overlooked is that the animals in some of these studies may not have been together long enough to form the strong attachment that is necessary before there can be a pronounced response to its disruption. This latter explanation is supported by a 1975 study done by Suomi and colleagues (1975). They separated five-year-old monkeys from their nuclear families. The nuclear family is a highly enriched living situation by laboratory standards including access to own mother, father, and siblings as well as other adult pairs with their respective offspring. This environment also includes large living spaces with shelves, wheels, and swings, both in the home cages and the play area.

When the monkeys were separated from the nuclear family and housed with familiar peers, strangers, or alone they showed different responses depending on the post-separation living condition. All of the animals were disturbed after separation but those that were housed with familiar animals or strangers returned to their usual behaviors in a short time. The monkeys that were housed with strangers were initially involved in early bouts of aggression but these quickly diminished and the monkeys rapidly established normal social relationships with each other. In sharp contrast to this, the monkeys that were housed alone showed the classic protest and despair response to separation. This study shows that the depressive response occurs in older monkeys as well as infants. It also seems to indicate that loss of affectional objects is the main influence in producing this response. When the monkeys were housed with animals they knew this reaction did not emerge. Likewise, when they were able to interact with other animals and establish new bonds to replace the old, the response was not as severe. By contrast, the animals that were housed alone and were deprived of any social interactions showed severe and persistant signs of despair. That this was mainly due to loss of affectional objects rather than the characteristics of the single cage, seems indicated by the fact that other monkeys in laboratories normally live a large part of their lives in single cages without developing this pronounced despair response.

Many aspects of the peer separation paradigm still need to be studied. Research examining variables such as sex, preexisting behavior patterns, species, and age needs to be done. The research that has been done in the mother-infant model needs to be replicated in the peer separation model, including the work on the biological measures such as sleep, heart rate, plasma cortisol levels, temperature, etc. Such studies will help to further define the peer separation model itself, and, by extending understanding of the similarities between the peer and the mother-infant models, open the ways to adding information to the etiology of human depression. These studies will not only help to define the peer-separation model itself, but help promote understanding of the similarities with the mother-infant model and thereby give clearer definition of the underlying mechanisms and factors common to both. Even though the peer-separation model may not be definitively established as a model for human depression, it is such a powerful response that it is important to study it in its own right.

METHODOLOGY FOR STUDYING DRUG EFFECTS ON PEER SEPARATION RESPONSE

Since the remainder of this chapter discusses how specific drugs interact with peer separation, a basic protocol will first be described. The underlying concept of this technique is that rhesus monkeys which are reared together in a group, without mothers, develop strong attachments to each other. Separation from the group causes individual monkeys to show a protest-despair response that is similar to that observed after mother-infant separation. The despair response is defined experimentally as an increase in huddling behavior accompanied by a simultaneous decrease in locomotor and exploratory behavior.

During separation, animals are typically housed in wire mesh cages in which they can see, hear, smell, but not touch other members of the group. This separation housing condition is identical to that provided for chronically single-caged housed monkeys. Each separation is bracketed by a preseparation social housing condition and a post-separation social reunion condition. Thus, in this paradigm, a three-week period consists of a preseparation social housing block, social separation, and a social reunion period. Separation blocks on drug are intercalated with placebo blocks so that active drug can be given on every other separation, i.e., once every six weeks. Two groups can be run simultaneously and the protocol staggered so that one group can serve as a placebo control for another group, and in later separation blocks, crossed over. Drug administration is typically begun on the first day of the preseparation socially housed week, continued over the weekend and ended on the last day of the separation week, i.e., 12 consecutive drug days. No drugs are administered during reunion weeks. The same drug or placebo administration protocol may be used in chronically single-cage housed animals, except of course, that the housing condition is not altered. The length of each block can be altered as appropriate for individual drugs.

Behavioral observations are typically recorded five days per week, mornings and afternoons, by trained observers who have been shown to reliably score the same behaviors with a correlation of -.90 using durations and absolute frequencies for operationally defined categories.

There is a small but growing literature which examines the effects of separation on a variety of neurobiological variables. This approach utilizes separation as the independent variable. By contrast the work summarized in this chapter examines the effects of several classes of agents on the response of rhesus monkeys to peer separation.

Some of the drugs which were studied have well-established biochemical effects in rodents on noradrenergic and/or serotonergic systems. Other agents were chosen because of their linkage with human depression on other levels (Kraemer et al., 1981; Kraemer and McKinney, 1979; Suomi et al., 1978).

ALPHA-METHYL-PARA-TYROSINE (AMPT)

AMPT, which blocks tyrosine hydroxylase and therefore, both norepinephrine and dopamine synthesis, potentiates the separation induced despair behavior in monkeys without having similar effects on preseparation behavior. If the dose is large enough, i.e., greater than 50 mg/Kg/day, sedative effects may appear and changes may be produced in both group-housed and separation conditions. However, at doses below 50 mg/Kg/day, there are no observable effects when the monkeys are group-housed, but there is a definite potentiation of the despair response in the separation housing condition. The relative stimulus deprivation condition of single-cage housing is not responsible for the AMPT effects observed when the animals are separated. Support for this statement comes from data which indicate that a dose of AMPT from two to 20 times the dose which effects socially separated animals is required before chronically single-caged animals are affected.

PARA-CHLORO-PHENYLALANINE (PCPA)

This inhibitor of serotonin synthesis, in doses ranging from 25-200 mg/Kg/day, did not have differential effects on behavior dependent on social housing conditions. At 100 mg/Kg/day, PCPA did increase huddling but this effect was observed equally in group and separation conditions. PCPA did not decrease locomotion at the same dose levels that increased huddling. Therefore, PCPA does not appear to potentiate the peer-separation despair response at any dose between 25-200 mg/Kg/day despite the fact that drug treatment did produce some despair-like behaviors. Doses up to 100 mg/Kg had no significant effects on the behavior of chronically single-cage housed monkeys.

FUSARIC ACID (FA)

Fusaric acid in doses between 10-20 mg/Kg/day had effects which were virtually opposite to those of AMPT. At 20 mg/Kg/day, FA increased locomotion and decreased huddling in the separation condition as compared to placebo without having these or other significant effects in the prior social housing condition. Doses of 20 mg/Kg/day had no significant effects on the behavior of chronically single-cage housed monkeys.

Thus, work with the three agents just reviewed strongly suggests an interaction between the noradrenergic system and separation in rhesus monkeys. Interestingly, enzymatic blockage at two different points in the synthetic pathway affects separation quite differently. When the synthesis of both dopamine and norepinephrine (NE) is blocked, the depressive response to separation is enhanced. However, when only NE is lowered by the use of FA, the depressive response to separation is prevented. Alteration of the serotonergic system by PCPA clearly affected the social behavior but did so equally in both separated and unseparated conditions. Thus, it had no interactive effect.

There are other drugs that are related to human depression and the effects of some of these on rhesus monkeys' response to separation will be summarized below.

AMPHETAMINE (AMPH)

AMPH has complex and multiple biochemical effects on neurotransmitter systems. From a clinical standpoint, it is not an effective antidepressant when used chronically. Some work has suggested that it may have short-term antidepressant properties. The latter is consistent with reports from those who use such agents outside of medical supervision ("the street").

Evidence from our initial studies indicates that amphetamine reduces despair behavior in socially-separated rhesus monkeys and in this respect, it resembles the effects of imipramine. However, unlike imipramine, amphetamine disrupts social behavior at doses that have antidepressant effects. This disruption takes the form of increased agonistic behaviors, which sometimes leads to serious wounding which necessitates restricting access of monkeys to each other.

This response to amphetamine occurred at the higher dose level of 3.0 mg/Kg. At the lower dose of 0.50 mg/Kg, the antidepressant effects were apparent. However, in general, antidepressant effects decreased and agonistic effects increased with increasing dosage. To the extent that AMPH acts like an antidepressant in this animal model, it does so at a great price, namely the breakdown of those behaviors which promote normal social cohesiveness.

IMIPRAMINE

Imipramine, of course, is a frequently used and clinically effective antidepressant. Since the results are so well documented in human depression, it seemed reasonable to attempt to test its effectiveness in our animal model of separation.

Two groups of young rhesus monkeys were subjected to repetitive peer separations. Midway through the procedure, one group was treated with imipramine hydrochloride (10 mg/Kg/day) and the other with a saline placebo. In comparison with placebo treatment, the imipramine treatment yielded significant behavioral improvement in a form and with a time course similar to that seen when the drug is given clinically to human depressives. That is, after a predictable period of time, beneficial effects became apparent, but when the drug was discontinued, the antidepressant effects also disappeared. These comparable results provide additional support for the validity of this kind of model.

ALCOHOL

The relationships between alcohol and depression in humans are important but complex. Humans sometimes drink in an effort to treat their depression; however, any euphoric effects are temporary and quickly evolve into depressive effects. On the other hand, the primary problem is sometimes alcohol abuse and the depressive effects are secondary. Another body of literature links separation and depression. Thus, we have three potentially interactive factors: alcohol, separation, and depression. The interaction among these factors is difficult to study in a controlled manner in humans. Recently, we have turned to animal models to examine how alcohol might influence the well-documented depressive response of rhesus monkeys to separation.

Alcohol was administered to juvenile rhesus monkeys at dose levels of 1, 2, and 3 g/Kg/day of 25% ethanol in a between-groups crossover design with each group alternately serving as a placebo control. The dose levels were selected to have large, moderate, and minimal acute effects on behavior on the basis of previous reports and on studies conducted in our laboratory.

At doses of 2-3 g/Kg/day, alcohol increased ataxia, produced blood alcohol levels between 165-319 mg/100 ml, and also increased the duration of huddling over that observed when the animals were separated from their peer group and treated with placebo. However, some of the effects of these dose levels of alcohol altered social behavior in the peer-housing condition as well as in the separation condition by producing a more severe despair response than was observed on placebo.

Alcohol at a low dose (1 g/Kg/day) decreased the occurrence of huddle and passive behaviors during separation by comparison to placebo and, therefore, resulted in a less severe despair response than observed on placebo. Specifically, alcohol at this dose significantly ameliorated the response to separation without effecting the despair-response in the group housing condition. Thus, alcohol has a biphasic response in our separation model. At a low dose, it behaved like an antidepressant, whereas at moderate to high doses, it was a depressant. Such a biphasic effect is not unknown to human alcohol researchers.

CONCLUSIONS

Separation is a very stressful event for many primate species as it is for humans. This is true whether the separation is from one's mother or from peers with whom one has formed close attachments.

Previous research has established that a variety of behavioral variables influence the response to separation. A smaller number of studies have documented some neurobiological consequences occurring as a result of separation. The studies summarized in this presentation document that a variety of pharmacological agents can have interactive effects on rhesus monkeys' response to peer separation. That is, they have effects on separation at doses that do not significantly affect the group social behavior.

Many forms of human psychopathology, depression in particular, are thought to involve an interaction between social stress, e.g., separation, and neurobiological parameters. Such interactions have been difficult to study in humans and in most instances a relationship is hypothesized but the specifics of how a social stress such as separation interacts with other variables have been lacking. The use of animal models as outlined in this presentation makes such studies possible.

We think that separation from an affectional object represents a valid and useful animal model for some aspects of depression. Other models may be more appropriate for different aspects of depression. However, one need not agree about the relationship of separation in nonhuman primates to human depression to appreciate the value of studying the separation reaction _per se_ and trying to understand it better.

In this context, such studies represent the use of a nonhuman primate model system to study interaction between two classes of events, specifically a social event such as separation and alteration of neurotransmitter systems. We are trying to develop pharmacological/neurobiological profiles of the separation response and hence, to a certain extent, be able to reason to the importance of certain systems in mediating attachment behavior.

Disruption of attachment bonds is influenced by a variety of factors, both social and biological in nature, and sometimes even these two are intertwined. One needs to view separation and depression from a nonlinear, nondeterministic perspective. No single class of variables is sufficient to understand such a complex psychobiological happening. For example, separation produces depression in only a certain percentage of individuals (human and animals). These are likely those who are vulnerable in some way, i.e., genetic, developmental, neurobiological, social, etc. Those who do not become depressed by separation from an affectional object alone, may, in their turn, become depressed at separation when an additional stress, such as abuse of alcohol, is added to the separation situation. This interactive view of separation-induced depression in human and nonhuman primates is strongly supported by the studies just discussed in this paper.

This type of research does not directly address the problem of determining the specific biochemical mechanisms whereby drug effects are translated into alterations in complex social behaviors. Although most of the drugs being used have predominant neurochemical effects (and we are documenting these in our studies), they each also have nonspecific effects. Thus, without direct mechanism studies, one cannot safely reason from the effects of drugs on behavior to underlying mechanisms of action. However, the continued development of behavioral-pharmacological analogies may pave the way for an in-depth investigation of shared neurobiochemical behavioral mechanisms across species.

ACKNOWLEDGEMENTS

The writing of this chapter was supported in part by Research Grant MH21892 from the National Institute of Mental Health, HD10570 from the National Institute of Child Health and Human Development, and by the Wisconsin Psychiatric Research Institute.

REFERENCES

Bowden, D.M. and McKinney, W.T. Behavioral effects of peer separation, isolation, and reunion on adolescent male rhesus monkeys. Developmental Psychobiology 5, 353-362 (1972).

Bowlby, J. Grief and mourning in infancy and early childhood. Psychoanalytic Study of the Child 15, 9-52 (1960).

Breese, G.R., Smith, R.D., Mueller, R.A., Howard, J.L., Prange, A.J., Jr., Lipton, MA., Young, L.D., McKinney, W.T., and Lewis, J.K. Induction of adrenal catcholamine synthesising enzymes following mother-infant separation. Nature New Biology 246, 94-96 (1973).

Coe, C.L., Mendoza, S.P., Smotherman, W.P., and Levine, S. Mother-infant attachment in the squirrel monkey: Adrenal response to separation. Behavioral Biology 22, 256-263 (1978).

Hinde, R.A. and Davies, L. Removing infant rhesus from mother for 13 days compared with removing mother from infant. J. Child Psychology and Psychiatry 13, 227-237 (1972).

Hinde, R.A. and McGinnis, L. Some factors influencing the effects of temporary mother-infant separation: Some experiments with rhesus monkeys. Psychological Medicine 7, 197-212 (1977).

Hinde, R.A. and Spencer-Booth, Y. Individual differences in the responses of rhesus monkeys to a period of separation from their mothers. J. Child Psychology and Psychiatry 11, 159-176 (1970).

Hinde, R.A. and Spencer-Booth, Y. Towards understanding individual differences in rhesus mother-infant interaction. Animal Behavior 19, 165-173 (1971).

Jensen, G.D. and Tolman, C.W. Mother-infant relationship in the monkey, Macaca nemestrina: The effect of brief separation and mother-infant specificity. J. Comp. Physiol. Psychology 55, 131-136 (1962).

Kaufman, I.C. and Stynes, A.J. Depression can be induced in a bonnet macaque infant. Psychosomatic Medicine 40, 71-75 (1978).

Kraemer, G.W. Effects of alterations in biogenic amine metabolism on the protest-despair response to peer separation in rhesus monkeys. Dissertation Abstracts Intl. 38, 6212A (1978).

Kraemer, G.W. and McKinney, W.T. Interactions of pharmacological agents which alter biogenic amine metabolism and depression: An analysis of contributing factors within a primate model of depression. J. Affective Disorders 1, 33-54 (1979).

Kraemer, G.W., Lin, D.H., Moran, E.C., and McKinney, W.T. Effects of alcohol on the despair response to peer separation in rhesus monkeys. Psychopharmacology 73, 307-310 (1981).

Levine, S., Coe, C.L., and Smotherman, S.P. Prolonged cortisol elevation in the infant squirrel monkey after reunion with mother. Physiology and Behavior 20, 7-10 (1978).

Lloyd, C. Life events and depressive disorder reviewed. Part II: Events or precipitating factors. Arch. Gen. Psychiatry 37, 541-548 (1980).

McKinney, W.T., Suomi, S.J., and Harlow, H.F. Repetitive peer separations of juvenile-age rhesus monkeys. Arch. Gen. Psychiatry 27, 200-203 (1972).

Reite, M. and Short, R.A. Nocturnal sleep in separated monkey infants. Arch. Gen. Psychiatry 35, 1247-1253 (1978).

Reite, M., Kaufman, I.C., Pauley, J.D., and Stynes, A.J. Depression in infant monkeys: Physiological correlates. Psychosomatic Medicine 36, 363-367 (1974).

Reite, M., Short, R., Kaufman, I.C., Stynes, A.J., and Pauley, J.D. Heart rate and body temperature in separated monkey infants. Biological Psychiatry 13, 91-105 (1978a).

Reite, M., Short, R., and Seiler, C. Physiological correlates of maternal separation in surrogate-reared infants: A study in altered attachment bonds. Developmental Psychobiology 11, 427-435 (1978b).

Robertson, J. Some responses of young children to loss of maternal care. Nursing Times 49, 382-386 (1955).

Robertson, J. and Bowlby, J. Responses of young children to separation from their mothers. Courrier, Centre International de l'Enfance 2, 131-142 (1952).

Robertson, J. and Robertson, J. Young children in brief separation. A fresh look. Psychoanalytic Study of the Child 26, 264-315 (1971).

Rosenblum, L.A. and Kaufman, I.C. Variations in infant development and response to maternal loss in monkeys. American Journal of Orthopsychiatry, 418-426 (1968).

Seay, B. and Harlow, H.F. Maternal separation in the rhesus monkey. J. Nerv. Ment. Dis. 140, 434-441 (1965).

Singh, M. Mother-infant separation in rhesus monkey living in natural environment. Primates 16, 471-476 (1975).

Singh, S.D. Effects of infant-infant separation of young monkeys in a free-ranging natural environment. Primates 18, 205-214 (1977).

Smotherman, W.P., Hunt, L.E., McGinnis, L.M., and Levine, S. Mother-infant separation in group-living rhesus macaques: A hormonal analysis. Developmental Psychobiology 12, 211-217 (1979).

Spitz, R.A. Anaclitic depression: An inquiry into the genesis of psychiatric conditions in early childhood, II. Psychoanalytic Study of the Child 2, 313-347 (1946).

Suomi, S.J. Repetitive peer separation of young monkeys: Effects of vertical chamber confinement during separations. J. Abnormal Psychology 81, 1-10 (1973).

Suomi, S.J. and Harlow, H.F. Production and alleviation of depressive behaviors in monkeys, in Psychopathology: Experimental Models, J.D. Maser and M.E.P Seligman, eds. Freeman, San Francisco (1977), pp. 131-173.

Suomi, S.J., Harlow, H.F., and Domek, C.J. Effect of repetitive infant-infant separation of young monkeys. J. Abnormal Psychology 76, 161-172 (1970).

Suomi, S.J., Eisele, C.D., Grady, S.A., and Harlow, H.F. Deparessive behavior in adult monkeys following separation from family environment. J. Abnormal Psychology 84, 576-578 (1975).

Suomi, S.J., Seaman, S.F., Lewis, J.K., DeLizio, R.D., and McKinney, W.T. Effects of imipramine treatment of separation-induced social disorders in rhesus monkeys. Arch. Gen. Psychiatry 35, 321-325 (1978).

Vogt, J.L. and Levine, S. Response of mother and infant squirrel monkeys to separation and disturbance. Physiology and Behavior 24, 829-832 (1980).

Chapter 11
Social Effects of Alterations in Brain Noradrenergic Function on Untreated Group Members

D.E. Redmond, Jr.

INTRODUCTION

The effects of experimental manipulations in some group members on the behaviors of untreated monkeys of the same social group have often been ignored. Either an entire group is manipulated, and "control" data are obtained from a parallel untreated group, or some members of the group are treated with the others serving as controls. In either case, potential data may be lost because of the multivariate effects of the behaviors of each monkey on each of the others, whether they are "treated" or "untreated." This paper will attempt to illustrate this point with new data from studies of an inhibitor of catecholamine synthesis, alpha-methyl-p-tyrosine (AMPT), which depletes dopamine and norepinephrine without immediate biochemical effects on other brain systems (Corrodi and Hanson, 1966; Moore, 1966; Spector et al., 1965; Udenfriend et al., 1966). Since the purpose of this paper is to illustrate the effects of this treatment on group members which are not treated, the specificity of the treatment is unimportant except for the fact that direct biochemical effects should be limited to the treated monkeys.

Although the principal point of this presentation may be so obvious as to be trivial, the values of social behavioral studies and their complexities are often ignored. The almost hopeless difficulty of linking primate behavior to brain function in a meaningfully specific way leads many neuroscientists to abandon studies of behavior entirely -- to slice and homogenize and isolate in search of the "truth" about behavior. Some, in their efforts to control all of the variables reduce the possible repertoire of primate behavior to that of the restrained pigeon while others, in their efforts to find functional relevance in natural behaviors, ignore the interactive complexities of social behaviors. Perhaps, therefore, in a symposium devoted to methods as well as to substance, a specific example will serve as a reminder of one aspect of the complexities that exist in the analysis of social behavioral data.

MATERIALS AND METHODS

Artificial social groups of feral <u>Macaca arctoides</u> were caged together at least four months before each study. All but one of the groups were adults with estimated ages from four to fifteen years, consisting of an older clearly dominant male, a younger male, and four females. To study effects in younger animals, one remaining group consisted of six laboratory-born females from one and one-half to two and one-half years of age. Two to four members of each group were treated in each study, with the remainder serving as controls. The monkeys were housed in indoor group cages measuring approximately 2 x 4 x 2 meters. All were fed the same commercial monkey chow during the daily observations and had free access to water.

The groups were observed at the same hour each day, five to six hours after some were drug treated, and behaviors were recorded on a check sheet by an experienced, but experimentally naive and blind observer. The following categories were recorded: social and self-grooming, threats, physical attacks, mounting (social and sexual), social perineal presentation, submissive responses (lipsmack, facial grimace, and locomotor retreat), huddling, and play (locomotor, exploratory or object related, and social or "rough and tumble") according to standard descriptions (Bertrand, 1969; Blurton-Jones and Trollope, 1968). A "total active" social behavior category included those behaviors in which the individual was the initiator or "giver" of the behavior. "Social responses" were the sum of behaviors in which the animal showed some response to the initiatives of another. All the behaviors were scored as frequencies, except for grooming, huddling, and play for which one unit was scored/30 seconds of duration, producing a modified frequency score. Social rank was determined daily by priority of access to food, cage positions, displacements, and agonistic encounters in which clear dominance or submission was established.

Alpha-methyl-p-tyrosine (250-300 mg/kg, Regis Chemical Company, Chicago) was mixed as a suspension in water and administered through a nasogatric tube twice daily to each treated animal for periods from 14 to 84 days. In repeated studies this method of drug administration was found to be without behavioral effects when only saline or water were administrated and behavior was observed five to six hours later. Some control animals, therefore, were not sham-intubated.

RESULTS

All monkeys treated with alpha-methyl-p-tyrosine completed the entire course of treatment without any sign of physical illness, toxicity, or weight loss. The compound produced changes in social behavior in every treated monkey regardless of age or sex, consistently replicating the findings of our earlier smaller study (Redmond et al., 1971). Representative social behaviors for 10 treated monkeys and 11 controls are shown in Figure 1. With the exception of presentation and social responses, alpha-methyl-p-tyrosine appeared to affect all types of social behaviors. Unchanged levels of presentation and social responses ("passive" responsive behaviors) supported the observation that the animals were not obtunded, and that their withdrawal from social activity was more a lack of initiative than a spatial withdrawal. Approximately one half of the treated animals fell at least one position in social rank, and one-half of these did not recover their positions within a few weeks of the end of treament. In all cases some change in behavior preceded the change in rank. All other

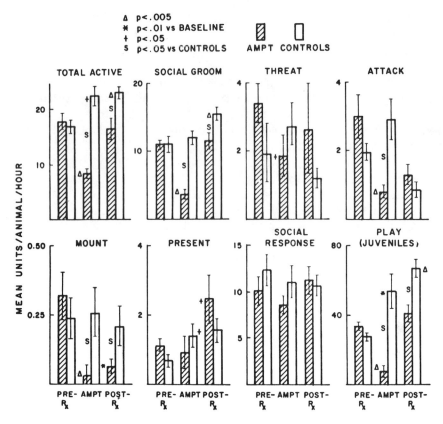

Figure 1. Social behaviors of ten monkeys treated with alpha-methyl-p-tyrosine compared with 11 untreated controls living in the same social groups. Three treated juveniles in another group, compared with three age and sex-matched controls, illustrate the effect on play behavior. Statistical tests were paired t within groups and Student's t between groups. None of the post-treatment activity levels of treated monkeys were different from their baselines.

behaviors did return to the pre-treatment levels after alpha-methyl-p-tyrosine was discontinued.

The "controls" showed significant changes also, almost always in opposite directions from the treated monkeys (See Figure 1). They increased their social ranks, showed more social behaviors, and increased a variety of specific behaviors. The controls increased "total active," "attack," and "play" behaviors relative to the treated monkeys and to their own baseline level.

DISCUSSION

This study illustrates the effect of treatment of some members of a social group on the behavior of other group members. Similar effects were found in studies of a catecholamine-specific neurotoxin, 6-hydroxydopamine (Bloom et al., 1969; Heikkila and Cohen, 1971; Redmond et al., 1973) and of the social effects of locus coeruleus lesions in the same species (Redmond, 1979; Redmond et al., 1976). This study therefore illustrates experimentally a theory which is often advanced concerning the importance of behaviors of other group members (usually family) on the behavior of individuals within a closed social group. These effects presumably occur because of alterations in interactive social behaviors, social signals, social ranks, and customary social roles of individuals within the group.

Although the possibility of this type of interaction is obvious from an analysis of the experimental design, it has a number of consequences which are not so apparent, and the alternatives are also not without disadvantages. First, it is not possible exactly to balance the social ranks of treated and control animals within a group. Second, tests of statistical significance are difficult. The usual conservative statistical test that might seem to be appropriate for these data, a two factor analysis of variance, with repeated measures on one factor (Winer, 1971), will often not detect real change in the treated animals because of large simultaneous changes in the controls. In fact this design clearly violates the assumption of independence necessary for the analysis to be valid. In order to circumvent this problem, separate t tests have been done for each treatment group, as shown in Figure 1. This model forces us, therefore, to utilize less powerful parametric or nonparametric statistical tests to determine significance.

An appropriate statistical test, however, will not solve the basic problem of the interaction. We cannot know whether the changes in behaviors of the control animals are secondary to the changes in treated monkeys, or vice versa. Perhaps both are interacting with some other unknown and uncontrolled variable. Sometimes the social signal, behavior, or variable responsible for some of the changes can be identified. An example of such an identification occurred during recent studies of the effects of naloxone in morphine-addicted monkeys. Three groups composed of a morphine-naloxone treated, a saline-treated, and an untreated monkey were caged in three balanced group cages. The experimental results were confounded when vocalizations by the naloxone-treated monkey undergoing "precipitated" morphine withdrawal in each cage produced similar vocalizations and other "withdrawal" behaviors in the controls. The hypothesis that these vocalizations and many of the behaviors were identical to those seen in response to fear or alarm was confirmed when the experiment was repeated with monkeys caged separately by treatments and also presented with a variety of other "fear-provoking" stimuli (Redmond et al., in preparation). It appeared that the vocalization by the treated monkeys in the previous experiment was serving as a "fear-stimulus" to elicit identical behaviors in the controls. However, in the latter situation, when the groups were separated by treatments, one cannot discount the possibility that the same behaviors were altered by the behaviors of their cagemates, who were either also vocalizing from naloxone-effects, or were relatively unresponsive from having received only morphine. In other words, treating all of the monkeys within a closed group with the same treatments does not eliminate the problem of behavioral interactions, but prevents confounding with treatments.

The use of a treated group and a separate, exactly matching, control group has other limitations as well. The large cage spaces required for monkeys, as well as the small number of animals which can justifiably be studied, often limits the employment of multiple treatment and control groups consisting of large numbers of subjects. Differences in cage size, design, and location, in addition to individual animal differences can also lead to significant differences between groups even prior to treatment, thereby making statistical inferences difficult.

An ideal, if uneconomic, solution would be to study only one subject monkey in a social group, compared with age-sex matched cagemates and with similar sham-treated monkeys in a matching separate but identical cage. If one were interested in normal group interactions, one would have to add a variety of animals of various ages and both sexes to each group. It is obvious that to study four or five treated subjects, one might have to study 50 to 60 monkeys. Realistically, this is not often possible; and let me be the first to admit that I have usually been unable to follow this advice myself, as these examples illustrate.

In conclusion, the rewards for paying attention to social interactive behaviors of primates are many and may include some information about the functional relevance of important brain phenomena (Kety, 1950). But interactive and social hierarchical effects on even solitary "non-social" behaviors must be considered. Pilot studies can often help to design the most definitive studies which are feasible in light of interactive behavioral effects. Finally, "conclusions" from studies purporting to link brain mechanisms to social behavior must be considered as hypotheses to be tested in more biologically specific ways, from molecular receptor sites for a single brain cell through circuits of thousands or billions of cells.

ACKNOWLEDGEMENTS

The original studies were undertaken with the collaboration of Drs. J.W. Maas and A. Kling. Drs. C. W. Graham, R. Gauen, F. Schlemmer and M.O.J. Justic were observers and provided valuable technical assistance. Partial support was provided by the State of Connecticut, and by grants MH 31176, MH 24607, and a Research Career Scientist Development Award KO-DA00075 from NIDA.

REFERENCES

Bertrand, M. The behavioral repertoire of the stumptail macaque. <u>Bibliotheca Primatologica, No. 11</u>. Karger, Basel/New York (1969).

Bloom, F.E., Groppetti, S.A., and Costa, A.R. Lesions of central norepinephrine terminals with 6-OH-dopamine: Biochemistry and fine structure. Science 166, 1284-1286 (1969).
Blurton-Jones, N.G. and Trollope, J.L. Social behavior of stump-tailed macaques in captivity. Primates 9, 365-294 (1968).
Corrodi, H. and Hanson, L.C.F. Central effects of an inhibitor of tyrosine hydroxylation. Psychopharmacologia 10, 116-125 (1966).
Heikkila, R. and Cohen, G. Inhibition of biogenic amine uptake by hydrogen peroxide: a mechanism for toxic effects of 6-hydroxydopamine. Science 172, 1257-1258 (1971).
Kety, S.S. A biologist examines the mind and behavior. Science 132, 1961 (1950).
Moore, K.E. Effects of alpha-methyl tyrosine on brain catecholamine and conditioned behavior in guinea pigs. Life Sci, 5, 55-65 (1966).
Redmond, D.E., Jr. The effects of destruction of the locus coeruleus on nonhuman primate behaviors. Psychopharmacology Bulletin 15, 26-27 (1979).
Redmond, D.E., Jr., Maas, J.W., Kling, A., Graham, C.W., and Dekirmenjian, H. Social behavior of monkeys selectively depleted of monoamines. Science 174, 428-431 (1971).
Redmond, D.E., Jr., Hinrichs, R.L., Maas, J.W., and Kling, A. Behavior of free-ranging macaques after intraventricular 6-hydroxydopamine. Science 181, 1256-1258 (1973).
Redmond, D.E., Jr., Huang, Y.H., Snyder, D.R., and Maas, J.W. Behavioral changes following lesions of the locus coeruleus in the stumptail monkey (Macaca arctoides). Presented at the 5th Annual Meeting of the Society for Neuroscience, Toronto, Canada (1976).
Redmond, D.E., Jr., Stogin, J.M., and Leahy, D.J. Similar behavioral effects of morphine abstinence and natural fear-inducing stimuli in C. aethiops monkeys (in preparation).
Spector, S. Sjoerdsma, A., and Udenfriend, S. Blockade of endogenous norepinephrine synthesis by alpha-methyl tyrosine, an inhibitor of tyrosine hydroxylase. J. Pharmacol. Exp. Therap. 147, 86-95 (1965).

Udenfriend, S., Zaltman-Nirenberg, P., and Gordon, R. Evaluation of the biochemical effects produced in vivo by inhibitors of the three enzymes involved in norepinephrine biosynthesis. *Molec. Phramacol.* 2, 95 (1966).

Winer, B.J. *Statistical Principles in Experimental Design*, 2nd Ed. McGraw-Hill, New York (1971), pp. 262-276.

Chapter 12
Strategic Psychopharmacotherapy: The Therapeutic Use of Medication in Family Systems

L. Engel and T.G. Bidder

INTRODUCTION

This paper deals with the interconnections of three very complicated processes: internal feeling state (mood); interpersonal distance (proximity); and biological processes (psychobiology). Their interrelationships are discussed by examining the effects of a psychopharmacological agent which resulted in the improvement of an identified patient's severe behavioral disorder when it was administered to another member of his family.

A model is proposed to deal with the interdependency of psychobiology, mood, and interpersonal proximity in transactions between individuals. It is used to analyze the family-wide behavioral effects, as described in a Case Report, which occurred when a psychotropic drug was administered to a family member other than the identified child patient. In addition, there is a discussion of how concepts and principles derived from several seemingly disparate disciplines can be used to explain how a drug could have altered the biopsychosocial balance existing in this troubled family with a consequent modification of the behavior of all its members.

Two concepts will receive special emphasis because of their implications for a more effective approach to the treatment of behavioral disorders. First is the idea that an individual family member functions as an inextricable part of the interpersonal relationship system operative in this family group (Ackerman, 1958). This is in contrast to the generally held view in which the individual member is considered primarily to be an autonomous and separate functional unit which happens to be located in a family group.

The second concept is that a psychopharmacological agent can exert actions external to the individual who receives the medication. This idea seemingly contradicts traditional pharmacological thinking which views drug actions as being confined to the symptomatically identified patient to whom the agent is administered and in whom therapeutic changes are subsequently expected.

The Model

Human individual and social behavior can be understood more specifically in terms of the interplay between the intrapersonal processes of psychobiology, mood (internal feeling state), and the interpersonal process of proximity (see Fig. 1). These three interactive processes have a vectorial equilibrium quality: changes in one are transduced to each of the other two within and between significantly interrelated individuals. This interplay is the basis for the complex homeostatic balance which is maintained between interrelated individuals. It exists, in a constantly changing state, during each moment of human interaction with shifts occurring in response to changes within and between individual members so as to maintain the complex homeostatic balance necessary for the group's survival.

One additional interactive factor needs to be mentioned: context. This is one social definition of a given setting which determines the meaning for, or the impact on, mood and psychobiological subsystems of the many possible spatial actions that individual group members may take. For example, an intimate context permits greater expectations of closeness and emotionality than does a business meeting. It is through context that the developmental and historical evolution of the individual's cognitive framework proceeds and operates.

Before presenting and analyzing the Case Report, some brief statements about psychobiology, mood, and proximity will be made to provide some

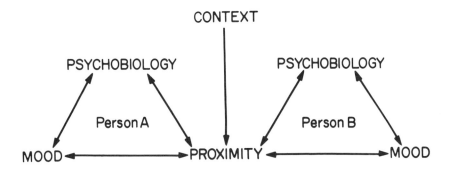

Figure 1. Interplay between psychobiology, mood, and proximity in a dyadic relationship.

perspective for readers who are unfamiliar with these terms.

Psychobiology

Psychobiology refers to the myriad of fundamental biophysical, biochemical, and physiological processes which constitute the basis of the human organism's functioning. These fundamental biological processes must maintain a rather narrow range of activity if normal functioning is to be preserved. For this purpose the living organism possesses powerful buffering, adaptive, and homeostatic mechanisms for preserving this functional competence in the face of a constant barrage of disruptive influences.

While the activities of these individual intra-cellular and inter-cellular processes largely proceed below the level of the individual's awareness, their collective functioning is experienced in the realm of what is termed the "internal feeling state" (IFS). Since mood or affect is a clinically important component of the IFS, the term mood is used interchangeably with IFS throughout the remainder of this discussion. Derangements in psychobiological processes which exceed an individual's buffering and homeostatic capacities are frequently experienced as dysphoric mood.

Mood

The term mood is used in this paper to designate one of the major and universally experienced qualitative modes of the IFS. Mood can be conceived of as a monitor (Pribram, 1970) which serves to alert the individual to the equilibrium and dysequilibrium resulting from the successes and failures of motivated action (Beck, 1976). Mood states which exceed some highly individualistic threshold values signal a need for the activation of a wide variety of adjustments designed to decrease the dysphoric mood.

Two of these corrective adjustments are of special interest here: alteration of interpersonal distance and affectual bonding. The latter term, which is attributable to Bowlby (1969), refers to the important need to attain or retain proximity to some special individual who is usually conceived to be wiser and/or stronger. The need for affectional bonding, though most marked in the very young, is retained throughout life.

Proximity

The need to establish a degree of closeness to, or distance from, significant other persons appears to characterize all humans (Bowlby, 1969). For any given relationship, the level of closeness or distance is not fixed but varies dynamically to provide a tolerable balance between the positive and the negative aspects of intimacy and separateness for each of the individuals involved. When, for whatever complex reasons, the comfort range for proximity is significantly exceeded or fallen short of within a given context, the involved individuals will attempt to compensate (Patterson, 1976) in order to minimize or eliminate the discomfort (Argyle and Dean, 1965). A major method for accomplishing this adjustment is an alteration of proximity.

Interrelationships of Psychobiology, Mood, and Proximity

Psychobiology and Mood

The interrelationships between these two processes are frequently striking. Thus in the affective disorders, severely dysphoric states are are associated with a wide variety of neurohormonal, neurophysiological, and neuropharmacological derangements (Carroll, 1980). Of special significance is the capacity of appropriate pharmacotherapy to correct the dysphoric mood and, _pari passu_, the associated psychobiological abnormalities.

Mood and Proximity

The degree and type of closeness required and tolerated is related to an individual's IFS. When a person experiences a sense of emptiness, i.e., a dysphoric IFS, there will be a physical and emotional movement toward some significant other in an attempt to "fill themselves up with the other," i.e., modify the dysphoric IFS by changing interpersonal proximity (Fogarty, 1978). Libowitz and Klein (1979) have observed that a particular group of individuals tend to modify their dysphoric moods "...by the use of others as mood adjusting agents." Bowlby (1969) has also looked at the relationship between specific types of interpersonal relationships and an individual's mood:

"Many of the most intense emotions arise during the formation, maintenance, the disruption and the renewal of attachment relationships. The formation of a bond is described as falling in love, maintaining a bond as loving someone, and losing a partner as grieving over someone. Similarly, threat of loss arouses anxiety and actual loss gives rise to sorrow; while each of these situations is likely to arouse anger. The unchallenged maintenance of a bond is experienced as a source of security and the renewal of a bond as a source of joy. Because such emotions are usually a reflection of the state of a person's affectional bonds, the psychology and psychopathology of emotion is found to be in large the psychology and psychopathology of affectional bonds." (Bowlby, 1977)

PSYCHOPHARMACOLOGICAL PERSPECTIVES

Psychopharmacological agents have potent effects on the above-mentioned interrelationships between psychobiology, mood, and proximity. Furthermore, as will be apparent from the Case Report which follows, they can produce significant systems-wide behavioral changes in a family group. For these reasons, certain general statements about drug actions need to be made at this point; other characteristics of drug action will be mentioned in a later section because of their usefulness in trying to understand the behavioral changes which were observed to occur when medication was introduced into a family system.

Most of the scientific knowledge that we possess about drugs and their actions has been obtained from studies of animal and/or human individuals and on cellular and subcellular preparations from these organisms (Gilman et al., 1980). While some of this knowledge is clearly related to the systems effect of drugs, certain other actions, though, have not been given much attention with respect to their possible relevance to behavioral studies. It is this latter group of drug characteristics which will now be discussed.

Drugs Modify Existing, Ongoing Psychobiological Processes Rather Than Creating New Actions

It is becoming clear that dysfunctional psychobiological processes which characterize certain psychiatric and behavioral disorders are corrected by therapeutically effective modalities, e.g., receptor abnormalities in certain depressions which are corrected with antidepressant medications and electroconvulsive therapy (Chiodo and Antelman, 1980; Creese and Sibley, 1981; Gillespie et al., 1979; Sulcer, 1979). In the case report which follows the response of the child's father to phenelzine was primarily the result of alteration of dysfunctional psychobiological processes which were a contributing factor in the genesis of his clinical state.

Certain characteristics of psychobiological processes are reflected in the actions of drugs on behavior. One of these is the periodic variations in the concentrations of certain body constituents, e.g., diurnal variations in serum cortisol, and in psychopathological states, e.g., intra-diem fluctuations in the intensity of depression (Carroll, 1980). Another example of this chronotropic property is seen in the different responses to a drug which occur depending on the time of day at which it is administered (Sermons et al., 1980).

Drugs Have Multiple Actions

Drugs rarely possess only a single action; an agent such as phenelzine, the monoamine oxidase inhibitor (MAOI) antidepressant used in the case example, has multiple effects (Gilman et al., 1980). These include antidepressant, antihypertensive, anti-anginal effects as well as microsomal enzyme-blocking, and monoamine oxidase-inhibiting actions.

Some of these actions of phenelzine are therapeutic while others are non-therapeutic and can be productive of adverse effects. What precise role the therapeutic/non-therapeutic actions of phenelzine play in its positive/negative systems effects is not clear.

Non-Linear Dose-Response Relationships

The usual and expected relationship between the dose of a drug and the effect it produces is linear. However, it is likely that a non-linear dose-response relationship is more common than is presently recognized. Some drugs exert a stimulatory action within one dose range and an inhibitory action at other, higher doses (Besser et al., 1980).

Non-linear dose-response relationships can result from the multiple actions of a drug--each of which has its individual relationship to the dose administered. The net pharmacological effect at any given dose or biophase concentration is the resultant--the algebraic sum--of each of these individual dose-response relationships. This can result in one effect being manifested at a certain dose while a different action will be predominant at another dose. Particularly with drugs acting on the CNS, certain brain structures will be responsive at one dose while others will be activated at other doses. Finally, the type of drug-receptor interaction may be of one type, e.g., stimulatory, at a certain biophase concentration and of a different type, e.g., inhibitory, at another. The interplay of these effects, perhaps, accounts for the fact that the profile of certain psychopharmacological agents' effects--the net pharmacological result of the therapeutic and non-therapeutic actions--sometimes shows a curvilinear dose-response relationship (Åsberg et al., 1971; Magliozzi et al., 1981).

Pulse (Hit and Run) Drug Effects

Generally, the physical presence of a drug in the biophase surrounding a receptor is considered to be a sine qua non for continued pharmacological action. However, there are reports of long-lasting drug effects occurring with a single administration of drugs which are comparable to those occurring with repeated daily doses (Chiodo and Antelman, 1980). Presumably in these cases, the drug's actions (e.g., receptor sensitivity modification, irreversible enzyme inactivation) are consummated with the first dose(s) and its continued presence in the region of the target structure is not necessary for a persisting pharmacological effect.

It is possible that the longitudinal changes in behavior associated with the administration of certain psychotropic drugs, e.g., the 10-day latent or lag period before full therapeutic effects occur in individuals receiving antidepressant drugs and of lithium, are a manifestation of temporal-lag, hit-and-run effects. A similar phenomenon, described in the Case Report, could be explained by time-lag processes: critical changes were observed at three weeks and at three months after the start of the psychopharmacotherapy.

Time-Related Changes in Drug Actions

Another type of longitudinal change in a drug's action is the familiar phenomenon of tolerance (Gilman et al., 1980). As tolerance develops, the intensity of one or more of the pharmacological actions present initially will progressively diminish or disappear unless the dose is increased. The development of tolerance is frequent with one or more of the actions of psychopharmacological agents.

Antidysphoric Strategies of Everyday Life

Although not a pharmacological action in the usual sense, human drug-taking activities are noteworthy: the selectivity involved in choosing a particular agent and the strength of the need--the compulsion--to continue to take it. The preferred individual use of alcohol, cigarettes, narcotics, or benzodiazepines are well known as is the great difficulty in terminating this type of usage. Implicit in selecting a particular type of agent, out of the many available, is that it best performs some psychic function, i.e., produces some desired alteration in the individual's IFS. The choice of the agent may provide some clues as to the type of dysphoric state which is being corrected as well as to the process by which this effect is being produced. Thus, amphetamine may be used to self-treat depressive dysphoria and to do so by a direct action or/and blocking the re-uptake of norepinephrine by adrenergic neurons or/and, when used regularly and continuously, by inhibiting monoamine oxidase.

Apparently, a great deal of human activity is devoted to the alteration of dysphoric states. A variety of modalities is used for this purpose: chemical agents, food substances, physical activities, and interpersonal relationships.

CASE REPORT: CLINICAL OBSERVATIONS AND INTERPRETIVE ANNOTATIONS

(1) Clinical Observation

The family consisted of father, a needy anxious salesman, mother, an apparently capable and reasonable housewife, and a ten-year-old boy who presented as the identified patient. He had been evaluated at a university medical center and diagnosed as being a "hyperactive" child. He had been treated with progressively increasing doses of methylphenidate (Ritalin) over the three-year period prior to our first contact with him. During this period there had occurred a progressive and significant deterioration of his behavior and in the behavior of the family in general. The child was experiencing many serious difficulties at home, his school performance and behavior were poor, and he had no significant peer relationships. Medication had been stopped prior to our initial contact with him due to the lack of a sustained therapeutic response.

(1) Interpretive Annotation

The "Hyperkinetic Syndrome" is actually a multifaceted complex of behavior patterns and physiological concomitants (Whalen and Henker, 1980), not all of which are responsive to psychopharmacological intervention. Instead of viewing the child as the identified patient, his behavior can be viewed as a stabilizing factor in the family (Barragan, 1976). All attempts to remove his behavioral problems via psychotherapy or through the use of medication (Ritalin) proved unsuccessful and, in fact, had resulted in an escalation of the family conflict. This is understandable: any treatment strategy which did not recognize the child's vital homeostatic role in his family would be pushing against the family's homeostatic processes (Jackson, 1968).

(2) Clinical Observation

The child's behavioral problems, as well as his designated patient status, were conceptualized as a manifestation of a family-wide conflict which included a hostile and over-close relationship between the parents. Father would constantly express his frustration at not getting his emotional needs met and mother would respond by distancing and seemed unable to understand father's distress. This pattern would rapidly escalate to a screaming argument with father chasing mother and mother moving away, which only resulted in father redoubling his efforts for closeness.

(2) Interpretive Annotation

The relationship between each parent's mood and the degree to which they are able to satisfy their individual requirements for optimal proximity are illustrated in Figure 2b.

In the context of this relationship, father is playing the role of "pursuer" (Fogarty, 1978), and has the smaller relative proximity requirements.

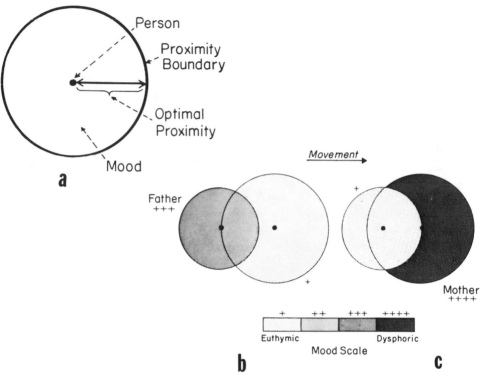

Figure 2. a - Notation used in diagrams; b - Parents' interrelationship at start of psychopharmacotherapy; c - Parents' interrelationship meeting father's optimal proximity-mood needs.

Father's movement towards mother is an attempt to reach a comfortable degree of closeness. Father's requirements for closeness, and his associated comfortable/euthymic mood, are represented in Figure 2b. Figure 2b represents a potential configuration which shows the optimal proximity-mood relationship for father which would result in a decrease in his dysphoria and terminate the activity of his attachment system.

> "Proximity-seeking attachment behavior that is not terminated increases in intensity and usually presents a picture of angry protest, depression, despair or any one of a number of defensive manoeuvres principally directed towards those who may be expected to offer the

> kind of interaction that will terminate proximity-seeking behavior." (Heard, 1978)

Therefore, father's dysphoric mood, in the context of this relationship, is the result of his inability to achieve an euthymic proximity relative to mother. Mother responds to father's attempts to get closer by moving away, maintaining her optimal proximity and, therefore, her euthymic moood. Father becomes increasingly frustrated and dysphoric as he is prevented from achieving the necessary closeness by mother's competing proximity requirements. The result is a continuously escalating conflict as both parents attempt to get all their needs for emotional stability and closeness exclusively from each other. This process has been described as fusion (Bowen, 1968). These fused relationships buffer the IFS of the individual members by keeping the interpersonal proximity within a tolerable range and accommodate to their differing proximity requirements. Accommodation occurs inter- and intra-personally through feedback modification of the interconnected processes of mood--proximity--psychobiology (Figure 3).

The tolerance ranges for proximity in this context are influenced by psychobiological factors. Psychobiologic factors influence the thresholds for the tolerance of dysphoric IFS and thereby determine the point at which mood-driven proximity-seeking attachment behavior will be evoked or terminated.

(3) Clinical Observation

Whenever the parental conflict reached an explosive point, the child's behavioral problems would escalate so as to monopolize the family's attention. This shift resulted in the parents now arguing about what to do with the child instead of arguing about their relationship. The emotional distress experienced by each family member was only temporarily reduced by the alteration between spousal conflict and parental concern over the child's acting-out. Tension within and between the family members continued to escalate unremittingly.

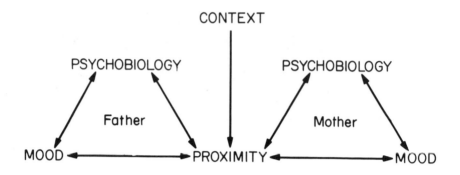

Figure 3. Interplay between mood, proximity, and psychobiology in relation to father and mother.

(3) <u>Interpretive Annotation</u>

Family systems' dysfunctions have been implicated as potential causal factors in a variety of childhood behavioral disorders (Stachowia, 1968; Bell and Vogel, 1960). The child's acting out appears to play a homeostatic role by flaring up precisely at the times when the parental conflict is nearing an out-of-control level, and the threat of the parents separating becomes a real possibility. The child's acting out serves to divert, displace, and distract the parents from their own conflict to his behavioral problems (arrows in Figure 4).

In this way the child is able to modulate the amount of interpersonal stress between the parents. By converting the stress between the parents into a concern over the problems of the child, the family's integrity was protected. This represents an example

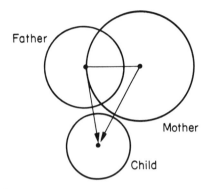

Figure 4. Triangulation involving family members. Arrows represent displacement of parental conflict onto child.

of the stabilizing processes of triangulation (Bowen, 1978). The trade-off for the rigid stability afforded by the triangular process is that the child is relegated to a highly dysphoric position (Figure 5).

At this too-distant position, the child's attachment system is in an activated state with the accompanying anxious, depressed escalating behavior. Due to the parental fusion he is constrained from establishing a degree of proximity which would then terminate the attachment system activation and the associated acting-out behavior. The child's role in the family made it necessary for him to stay at home and maintain a constant vigil on his parents. This is one explanation for his lack of peer interaction and of play. Another applicable explanation involves the concept of "holding the situation" (Winnicott, 1958). Under circumstances where children find themselves in an unsafe parental

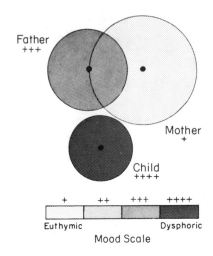

Figure 5. Relationship between proximity and mood in the triangled family.

environment, and do not have a stable base from which to explore away from home, they must therefore remain close to home.

(4) <u>Clinical Observation</u>

The family was treated for a period of four months as a unit without the use of medication. A variety of psychotherapeutic strategies were utilized with little success in trying to help the parents to contain their conflicts within the confines of their relationship and to deal effectively with their parenting function. The therapeutic strategy was one of decreasing the relationship intensity and increasing the differentiation of the parents so as to allow the child to be removed from his role as the official family problem. However, all attempts to decrease this over-involvement and fusion proved extremely difficult for the couple to tolerate, and resulted in an increase in their already discomforting anxiety. Alternate strategies which would make these changes more tolerable were therefore sought.

(4) **Interpretive Annotation**

Psychotherapeutic meetings with all the family members were undertaken as a means of restructuring the systemic imbalance in the family relationships. The goal of the therapy was to shift the family's triangular process to three independently-functioning dyads. Focusing the parental conflict back onto the couple, and thereby removing it from the child, would allow for independent dyadic interactions. This shift in family structure would allow the child to establish a close enough relationship with either of his parents and therefore remove him from the role as a central focus of the family's attention and apparent concern.

It was evident that alterations in the social-emotional proximities of the family members, instituted with the aim of prompting healthy differentiation, were insufficient to unlock the mood and psychobiological systems that appear operative in maintaining the triangle and stabilizing the family. It was very difficult for the parents to tolerate any separation because their fused relationship was the source of their relatively stable mood. Another characteristic of the process of triangulation is that the sum of the distances between the three members is a constant; therefore movement of any member toward or away from any other member will result in a change in position of the other members. Any increase in distance between the parents, which would be required for the child to move closer to either parent and therefore to a less dysphoric proximity, was vigorously resisted. The resistance to increasing the distance between the parents occurred because such a mood would shift them out of their narrowly tolerable mood-proximity range and result in increased dysphoria. The triangle can thus be viewed as a pathological response to the requirement that the family serve as a "distance regulating mechanism" (Kantor, 1975).

"The function of this relationship can be seen as protective and educative; it also has a part to play in regulating the physical distance and the amount of social interaction that takes place between people who know each other well. From this standpoint a family may be seen as a homeostatic system of relationships between a number of individuals at different stages of development. The system is held together by...goals aimed at adequate termination of proximity-seeking attachment behavior." (Heard, 1978)

(5) Clinical Observation

The use of specific types of medication, and the immediate and titratable effects on affective reactivity was considered as a possible way of enabling the parents to tolerate the increase in dysphoria which would be the result of decreasing the fusion in their relationship. As a means of assessing the parents' possible responsivity to medication, we administered a MAOI-Responsivity Checklist to both mother and father (Liebowitz and Klein, 1978). The mother and father met 75% and 95% of the criteria, respectively. These scores indicated probable responsivity to an MAOI antidepressant: while the father did not manifest the florid classical signs and symptoms of an endogenous depression, his Checklist responses and his clinical state suggested that he might respond to an MAOI antidepressant in a way that could have a salutory effect on family's interactional processes--especially those involving the parental dyad. The father's clinical state included: extreme jealousy, emotional lability, and marked sensitivity to perceived rejection. This, together with his willingness to take the medication, prompted us to undertake a therapeutic trial with phenelzine. In carrying out this trial we were, in effect, selecting the father as the best available portal of entry for introducing the medication into the family system.

(5) <u>Interpretive Annotation</u>

It was observed that both parents manifested signs of high affective reactivity, including rejection sensitivity, extreme jealousy, and sensitivity to criticism. It was also evident that their moods, especially father's, were very reactive to slight changes in the quality of the couple's interaction.

Research by Libowitz and Klein has focused on the relationship between the above sensitivities and an affective disorder which they have termed "Hysteroid Dysphoria".

> "The hallmark of the disorder is an extreme intolerance of personal rejection, with a particular vulnerability to loss of romantic attachment... the primary defect of hysteroid dysphorics is their affective vulnerability...these patients have a pathologically heightened emotional reactivity to approval or disapproval...other aspects of their behavior are expressions of this defect or attempts to compensate for it. Their emotional reasoning is seen as the domination of thinking and judgement by mood; their sexual provocativeness and manipulativeness as an attempt to elicit attention and approval; their lack of depth and consistency in personal relationships as secondary to the use of others as mood adjusting agents." (Liebowitz and Klein, 1979)

These authors have suggested that a MAOI antidepressant can alter an individual's reactive threshold. Based on our hypothesis that the processes of mood, psychobiology, and proximity are interconnected and operate within and between interrelated persons (Figure 6), a drug which can alter the threshold in the proximity-mood system within an individual family member will have its effects distributed throughout the whole family.

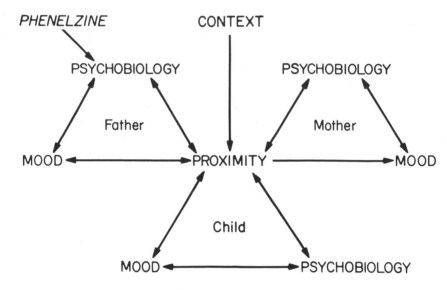

Figure 6. Interrelationship between the processes of mood, proximity, and psychobiology in this three-member family given phenelzine.

Given this hypothesis, each member of a family represents a potential portal for intervention with a psychopharmacological agent such that its effects will be distributed throughout the entire family. It was hoped that MAOI antidepressant administration might provide the kind of intervention that appeared necessary to adjust the mood and psychobiological imbalances of this family and one which the family was unable to undertake by non-pharmacological means. Hopefully, the medication would enable the parents to tolerate a slight decrease in fusion by buffering any resulting increase in dysphoric mood. The choice of father was based on the fact that his dysphoric position and proximity requirements vis-a-vis his wife were generating a continuous level of parental conflict. By enabling him to tolerate less closeness without increasing his dysphoria, the tension between the parents could be reduced so as to allow shifts in the functioning of the family that would not otherwise have been possible. The intent was not to relocate the "problem" in either parent, but rather to restructure the entire family.

(6) Clinical Observation

The following changes were observed after initiating psychopharmacotherapy in the father: (1) There was a significant reduction in the intensity of the symptomatic items on the father's MAOI Checklist, i.e., rejection sensitivity, extreme jealousy, and emotional lability, starting within three weeks and reaching a maximal response in four weeks; (2) two to three weeks after the medication was started father decreased his constant pursuit of mother and seemed more able to tolerate the reduced amount of emotional contact with her. The result was a dramatic reduction in the marital conflict. This, together with a decrease in the parental overcloseness, i.e., fusion, made it possible for the parents to spend more time in direct positive, in contrast to negative, contact with the child; (3) at about the same time as the marked reduction in parental conflict occurred (approximately two-three weeks after starting the father's medication trial), there was a significant reduction in the child's behavioral problems. The changes in the child's behavior included the establishment and maintenance of significant peer relationshps and the exhibition of a marked reduction in aggressive and disruptive behavior at home and in educational and recreational situations. Furthermore, projective assessments of the child's mood states, through the medium of art therapy, revealed a parallel improvement.

(6) Interpretive Annotation

The resulting shifts in mood, proximity, and psychobiology occurred throughout the whole of the family system. Father was able to tolerate a less fused and increasingly differentiated (Bowen, 1978) position vis-a-vis mother, i.e., improved mood, more distance, and greater flexibility, without increasing his dysphoria. As father and mother were able to funciton in a more differentiated way with each other, they no longer needed the child to stabilize their relationship. This decrease in fusion and increased distance between the parents made it possible for the child to move to a more stable proximity range with his parents, i.e., he was no longer kept at a dysphoric proximity. The result was termination of the activation of his attachment system and the associated acting out behavior.

Figure 7 shows how father's proximity requirements have changed due to the MAOI, and how the child is able to assume a closer proximity by the compensatory changes brought about by the family's triangular process. The child was now at a proximity such that his prior attachment system activation and associated behaviors were terminated. The new circumstances in the family provided a parenting environment which was adequate for the child to explore away from home (Heard, 1978). As a result, the child no longer needed to remain at home in order to monitor his parents' relationship; consequently, he was able to establish stable peer relationships for the first time. The entire family was modified, with changes observable in the behavior of each individual and in their interrelationships.

(7) Clinical Observation

As the father responded to the medication and his behavior towards his wife changed, the parental conflict diminished as the balance of dysfunction within the couple shifted. Rather than father losing control and mother stabilizing him, as was their previous pattern, he was now actually more stable than mother. The mother, however, continued her attempts to re-establish the old balance in the relationship by provoking and escalating conflicts. The father's new stability rendered these provocations unsuccessful. As a result of her inability to re-establish her prior dominant and controlling

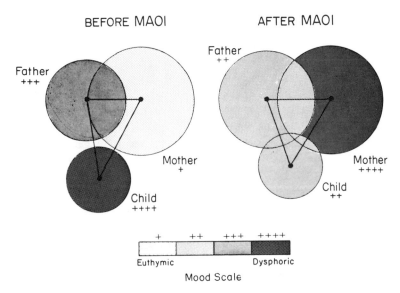

Figure 7. The relationship between changes in proximity and mood as influenced by MAOI antidepressant administration.

role, the mother became more and more depressed and for a time was suicidal. Father was able to tolerate his wife's pursuing as she attempted to re-establish the old balance, but was unable to tolerate her withdrawal and depression. As this intensified, he became increasingly anxious, jealous, and irritable. At this time mother was retested with the MAOI-Responsivity Checklist, and her score had changed from 75% to 90%. As mother withdrew, father's initially positive drug effect, i.e., decreased emotional lability and reactiveness, failed to be sustained. At approximately three months into father's treatment the positive drug effect had been lost, as mother was now more distant than father could tolerate. However, the child's improvement was maintained despite the deteriorating relationship between the parents.

(7) <u>Interpretive Annotation</u>

Father was now able to tolerate a decrease in proximity vis-a-vis mother and the child took up a new position closer to father. This shift resulted in mother being at a dysphoric proximity and unable to close to her prior optimal proximity (see Figure 7). Mother was unable to readjust so as to tolerate father's decreased fusion and increased distance from her. In an attempt to decrease her proximity-dependent dysphoria she became the "pursuer" in the relationship, manifesting extreme anxiety, protest, and depression and, for a time, suicidality. This incomplete re-balancing of the entire family system now placed mother in the "identified patient" role! She did not become "ill" but rather, as a result of a shift in the balance of the family, was held at a dysphoric proximity. The changes in mother's mood were evident in her behavior and also by the increase in her MAOI criteria score. This dramatically illustrates how context-dependent individual psychopathology can be. It is clear that mother's change in mood was linked to father's drug-induced psychobiological-proximity changes.

(8) <u>Clinical Observation</u>

After twelve weeks of psychopharmacotherapy, the father discontinued the phenelzine on his own initiative. Two major changes were observed to occur over the ensuing two weeks: the father's behavior, which had begun to get worse beginning at three months after the start of drug treatment, progressively deteriorated to the earlier chaotic pattern and the child's former symptoms rapidly reappeared. By the time that three drug-free weeks had elapsed, the child's behavioral improvements had been lost and he had returned to his previous dysfunctional behavior. During this period of change in the father and child, the mother became less depressed although her mood did not return to its pre-medication state.

(8) _Interpretive Annotation_

As the effects of father's changes were distributed throughout the family system, mother became dysphoric and escalated her behavior to re-establish the old homeostatic balance. Father was able to tolerate mother's escalation as long as she did not increase her distance beyond the new setting of father's activation threshold. Mother, faced with a more stable and distant husband and unable to reestablish the old balance, became anxious and depressed as a result of the activation of her attachment system. Father now found himself in a situation in which mother was even more withdrawn than before and at a distance which was greater than he could tolerate even with his drug-induced threshold elevation. In this new context, father's positive drug effect was lost, i.e., the medication was now unable to shift the activation threshold to the new extent required. In other words, the MAOI which had improved father's behavior beginning at three weeks into the trial apparently lost its effectiveness at approximately three months into treatment due to a shift in the context-dependent threshold beyond which activation could not be prevented.

The effectiveness of the MAOI on father can be understood in terms of the relationship between father's activation thresholds and the shifting dynamic balance within the family. When father was initially evaluated for possible MAOI responsiveness it was obvious that he was at dysphoric proximity, i.e., he was in a dyadic relationship in which he was unable to reach an optimal proximity and in a state of attachment system activation which was the result of exceeding his activation threshold. He was then treated with the medication, and for a time was much improved as manifested by a marked change in his behavior, i.e., less protest, despair, and anger. This could be understood in terms of a change in his activation threshold in that he was now able to tolerate the distance between himself and his wife which, before institution of the drug treatment, he had been unable to deal with and which kept him at a continuous level of attachment activation. When treated, father had a new threshold, but the threshold was contingent on the relationship between mother and father. Therefore any change in mother's behavior which involved alteration in interpersonal and emotional distance would have a direct effect on whether father's threshold was exceeded with the consequent activation of his attachement system.

Father reacted to the loss of the positive drug effect by discontinuing the medication and therefore removing the remaining psychobiological mood/proximity "buffer" that the MAOI had provided. It was inevitable that the family system would return to a pathological status if the drug-induced readjustment did not address all the levels of the family's dysfunction: mother's behavior and mood indicated a gap in the intervention. Shortly, the family was in fact once again immersed in the former pattern, and not surprisingly mother's mood substantially returned to her prior stable condition.

(9) Clinical Observation

This chaotic family situation and the child's severe behavioral relapse continued unrelentingly for approximately four months, at which time re-institution of therapy with lithium and phenelzine was promptly followed by an excellent therapeutic response in the father. The child showed significant improvement beginning two-three weeks after the onset of his father's behavioral changes, again reflecting a shifting of the emotional balance within the family. At approximately two months into this second drug treatment of the father, mother was moderately depressed but at this time did not manifest the suicidal feelings as she had during the last treatment period.

(9) Interpretive Annotation

The "setback" once again persisted until a psychopharmacological intervention was again applied. Father was then restarted on phenelzine at the same dose level as before. Again, systemic improvement was noted, as predicted, rather than simply a change in father's functioning. Bowen's observation that individuals with equivalent degrees of differentiation will form relationships sheds light on mother's behavioral responses to father's and child's changes (Bowen, 1978). Increasing father's differentiation through the use of medication resulted in an imbalance in the family and specifically in the parents' relationship. In this instance the major expressed imbalance shifted from the child to mother. The parents were able to function more as a couple after father was treated, as he was then not as affectively unstable as before. The result was that the conflict which required triangulation was decreased and more dyadic interaction occurred. The goal of detriangulating the family was not met but a more favorable balance was achieved, and the couple was able to function more and more as a dyad.

Mother's dysphoria in response to father's drug-induced behavioral changes as well as the change in her MAOI responsivity scores raise the possibility of treating both members of a couple simultaneously. This would allow mother in this case to parallel father's change in differentiation and further enhance the functioning of the couple. This raises additional questions to those already addressed in this discussion, and constitutes one of the areas under investigation at this time.

FURTHER CONSIDERATIONS

In what follows, the multidisciplinary material which has separately been discussed earlier in a topical manner is brought together for the purpose of trying to understand the behavioral changes which were observed to occur when medication was introduced into this family system. This will be done in terms of what we consider to be two of the most important and least discussed aspects of the actions of psychopharmacological agents: their context-dependence and their systems actions.

Context-Dependency

There is a growing awareness of the fact that the quantitative and/or qualitative aspects of a psychopharmacological agent's actions can be importantly influenced by the social and interpersonal context in which the individual recipient is embedded at the time of the administration of the drug. Certain representative animal and human observations which support this thesis will be cited.

The context-dependence of the behavioral actions of a number of drugs has been observed in several animal studies. Delgado and co-workers demonstrated that the behavioral effects of oral diazepam in a rhesus monkey varied with the animal's position in the social hierarchy (Delgado, 1976). The same animal showed changes in sleeping, alerting, and locomotion plus grooming behavior which differed, quantitatively, depending on whether it was in a dominant or submissive social position. When submissive, the animal was far more sensitive to the effects of diazepam on these behaviors than when it was socially dominant (Delgado, 1976). The effects of amphetamine on rhesus monkey behavior varied with the animal's position in its social group (Haber, 1977) and, in rats, differed depending on whether the animal was isolated or in groups (Haber, 1977). Sensitization or reverse tolerance to a variety of cocaine-induced behaviors occurred only when the drug was administered in a test cage (Post et al., 1971). The role of environmental context has also been shown in morphine tolerance (Siegel, 1976).

The context-dependency of psychopharmacological agents' behavioral actions has also been observed with humans. Earlier reference was made to the important influence of the biopsychosocial context on thresholds for mood, psychobiological processes, and proximity factors. The Case Report describes how the father's initial positive therapeutic response to the medication was lost after three months of psychopharmacotherapy as a result of the changes in the psychosocial context which enclosed him, i.e., the changes in interpersonal relationships. Finally, there is the frequently made observation that a psychotomimetic drug's effects can be pleasant in one social-interpersonal context and extremely disturbing in another.

Klein's observations on the activation and the release of specific affective states, e.g., the attachment system and the despair-depression system, are concerned with the consequences of exceeding the activation thresholds for these states (Klein, 1981). He has hypothesized that the positive action of antidepressants in the treatment of panic disorders occurs via modification of the psychobiological threshold processes resulting in the elevation of the activation thresholds for these specific affective states.

The existence of specific threshold levels is the expression of a complex interdependent process beginning with genetic predeterminants and extending to an individual's social context. In the Case Report we have discussed the fact that activation thresholds are also influenced by the social context. The social context's influence on activation thresholds will also influence the effect of a drug on these thresholds. The effect of a specific drug, in terms of its ability to raise the activation threshold, in this case for specific affective states, has been shown to be intimately linked to the biopsychosocial context of the father in the Case Report. The father's treatment with the MAOI, which had improved his behavior beginning at three weeks into the trial, apparently lost its effectiveness at approximately three months into treatment. This can be explained as being due to a shift in the context-dependent threshold beyond which activation of his attachment system could not be prevented.

Systemic Actions of Psychopharmacological Agents

Mention of the second important action of psychopharmacological agents -- their systems effects -- is virtually absent from the literature. In fact, the possibility that a drug is able to exert significant behavioral actions beyond the boundaries of the organism to which it is administered seemingly conflicts with generally accepted views of pharmacological mechanisms. Nevertheless, support for such extra-recipient drug actions has been obtained by McGuire in vervet monkey colonies, and is evident from the observations in the Case Report.

Raleigh and McGuire showed that the behaviors of untreated members of a vervet colony were influenced when individual animals received drug treatment (Raleigh and McGuire, 1980). When the dominant male received chronic treatment with the tryptophan hydroxylase inhibitor, parachlorophenylalanine, the frequency of grooming among non-treated, non-dominant group members dramatically and significantly decreased. In contrast, when a single non-dominant group member of the colony received similar treatment, no change in the grooming behavior of other, non-treated animals was observed.

Further support for the systems actions of a psychopharmacological agent comes from our observations, as recorded in the Case Report, that phenelzine produced profound behavioral effects on an entire family system when it was administered to a single member of the group--someone other than the individual who served in the role of a designated-identified patient. It is difficult to explain these observed family-wide behavioral changes without invoking drug actions which transcend the bodily boundaries of the individual who received the medication.

The observations recorded in the Case Report provide some additional information on the nature of the drug's systems action. It is clearly evident that a psychopharmacological agent introduced into a family via any single recipient inevitably produces systems-wide actions which, intended or not, are a mixture of positive effects, e.g., the child's improvement when father was treated with phenelzine, and negative reponses, e.g., the child's lack of improvement and the family's deterioration when he was treated with methylphenidate and his mother's symptomatic behavior when father was treated with phenelzine.

The child's lack of sustained improvement with the methylphenidate, while seemingly simply a failure to respond to the drug, can be understood, alternately, as a manifestation of the net result of the positive and negative effects evoked by the drug. As the symptom-bearing, designated-identified patient in this family, the child played a key homeostatic role in maintaining the emotional stability--albeit in a symptomatically psychopathological state. Through the triangulation fostered by his behavioral problems, the child served as an outlet or focal point for the family's collective tension and, thereby, helped ward off a serious deterioration in the parents' relationship. With this in mind, it is understandable why attempts to decrease the child's homeostatic patienthood with methylphenidate were countered by the seemingly paradoxical intensification in the family's conflicts. This escalation was countered by the child's remaining symptomatic. The ebb and flow between these countervailing forces were manifest symptomatically by the child's becoming unresponsive to the drug following an initial decrease in symptoms.

This final section of the discussion is concerned with the important yet difficult task of applying these concepts about the systems actions of psychopharmacological agents to the care of patients and families. Certain concepts which have been developed and discussed earlier are so essential for the clinical application of systems psychopharmacology that they can profitably be restated at this point. First is the recognition that the family group represents a single functional entity and that any seemingly individual behavioral disturbance is, in fact, indicative of a psychopathological process involving the entire group. Instead of being the sole locus of a disturbance, the most symptomatic member can be, instead, a selected barometer or indicator of family-wide difficulties (Bowen et al., 1961).

Related to this is the fact that a designated-identified patient, despite the obvious behavioral disturbance, may not be the best candidate for psychopharmacotherapeutic intervention. In fact, such an individual may be the worst candidate because his/her behavioral disturbance is playing a key role in maintaining a family in a precarious state of emotional equilibrium or stability (Jackson and Weakland, 1961). Finally, there needs to be recognition of the fact that introduction of a psychopharmacological agent into the system, albeit via a single member, inevitably produces family-wide actions which are positive, or beneficial, and negative, or adverse. Skill needs to be employed in using psychopharmacological strategies so as to maximize the positive actions and to minimize negative effects which can cause existing dysfunctional family patterns to become further entrenched and more rigid.

All families show some degree of resistance to changes necessary for modification of the dysfunctional patterns which are serving to maintain the integrity of the group. These dysfunctional patterns require a great amount of the family's resources to maintain with the result that there is a decrease in the family's capacity to deal with stress or, more importantly, to change. The dysfunctional patterns adopted to constrain movement and maintain acceptable proximity within these families must be very rigid to serve as adequate buffers against change and increased dysphoria. The large number of therapeutic failures encountered with these enmeshed and highly rigid families is understandable in terms of the difficulties involved in producing structural changes.

A family which is employing virtually all of its flexibility for emotional survival will have great difficulty in undertaking therapeutic changes which, initially, further increase the whole system's collective dysphoria. And therapists, not unexpectedly, encounter the resistance engendered by these powerful homeostatic processes buffering against further dysphoria.

It is with these difficult families that pharmacological intervention is most appropriate, in that the medication can increase the family's tolerance for change and therefore its flexibility to change. It is important to note that a psychopharmacological intervention with a family group requires only that the selected recipient be willing to take the medication. There need be no further behavioral or cognitive participation or cooperation on the part of the family to produce the type of structural changes described in this paper.

With these concepts in mind, is it now possible to address the question of putting them into practice via the conduct of a therapeutic trial of psychopharmacological intervention? When considering this type of therapeutic approach one needs to consider certain specific issues. Thus, is drug therapy appropriate, necessary, or even safe for a given family situation in contrast to other forms of interventions? Is it possible to select the most propitious group member to whom an appropriate psychopharmacological agent is administered? Which of the available drugs should be tried? How can the desired positive therapeutic actions of the selected agent on group and individual behaviors be maximized and the inevitable negative, adverse effects be minimized? The answers to these and related questions are the subject of continuing investigation.

EPILOGUE

> "The explicit philosophy of the general systems view is that reality is a tremendous hierarchical order of organized entities, leading, in a superposition of many levels, from physical and chemical to biologic and sociological systems. Unity of science is granted, not by a utopian reduction of all sciences to physics physics and chemistry, but by the structural uniformities of the different levels of reality." (Bertalanffy, 1968)

The systems viewpoint of human behavior espoused in this paper holds that living systems, from subcellular moeities to supra-individual collectives, embody a limited number of subsystems which essentially differ only in the degree of complexity as they participate in structural and

TABLE I

LEVEL OF COMPLEXITY	STRUCTURE	RECEPTOR	DRUG	SYSTEMIC EFFECTS PRODUCED
Sub-cellular	Hemoglobin molecule	Amino acid residues of $beta_2$-subunits	2,3-DPG*	Reduced oxygen affinity
Cellular	Skeletal muscle cell	Neuromuscular junction	Acetylcholine	Muscle contraction
Group	Family system	Drug recipient	Phenelzine	Behavioral changes

* - 2,3-diphosphoglycerate

processing functions carried on at each hierarchic level (Miller, 1978). This increase in complexity in ascending hierarchical levels involves structural and conformational elaboration which enables more complex functioning to be carried out. For example, in living systems chains of relatively simple amino acids form a folded and compact protein molecule whose unique spatial and biochemical properties are largely derived from the sequence of amino acids in the chain. As a result of this three-dimensional conformational pattern, certain unique biological functions are acquired by the protein macromolecule. These can include a specific and discriminative receptor capacity so as to interact with certain small molecules and, as a result, to transmit information throughout the macromolecule and thereby modify its biological behavior (Goodford, 1980). The same subsystems involved in this receptor macromolecule participate, albeit in much more complex conformational states, in functions of the cell, the organism, and in groups of individuals.

STRATEGIC PSYCHOPHARMACOLOGY   315

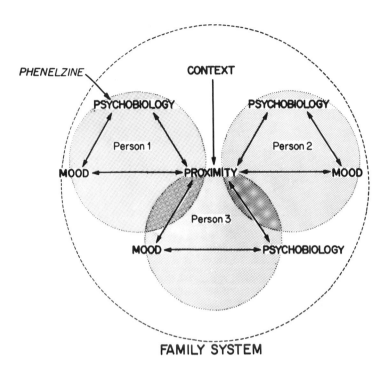

Figure 8. Pathways of conformational changes evoked by interaction of the drug with the family receptor.

One of the great achievements of modern biochemistry and pharmacology has been an elucidation of the molecular events which follow the relatively simple process in which a drug or other chemical agent specifically combines with some type of receptor to initiate a series of physiological or biochemical reactions in some complex biological structure. Table I presents three examples, at different levels of complexity, of the actions of chemical agents on complex biological systems.
In each of these examples, a drug or other chemical moeity produces conformational changes in the system which extend distal to the point at which it interacts with the receptor. Thus in the case of the hemoglobin molecule, combination of 2,3-DPG with certain amino acid residues of the two beta-2 subunits reduces the oxygen affinity of iron atoms located at a distance from the site of this combination. In the skeletal muscle cell, combination of

the neurotransmitter, acetylcholine, with a single neuromuscular junction or motor end-plate, causes distally-located myosin to rearrange its structure with the result that the entire cell contracts. With the family, contact of the drug, phenelzine, with a receptor--the drug recipient--produces changes in all members of the system. In all three examples, the intimate mechanisms by which the chemical agent or drug produces its system effects are unknown.

By analogy with the seemingly simpler examples, it is tempting to regard the individual recipient of the drug as the "receptor" for the family group system.

By further analogy, the "family receptor," properly chosen, will have the requisite specificity. That is to say, it will have the characteristics such that, upon being engaged by the drug, the agent's chemical or pharmacological information will be transduced into the desired perturbations of the family system.

ACKNOWLEDGEMENTS

The authors gratefuly acknowledge the help given by Patricia Derby, B.A., Marjorie Mattucci, and Julian Norton-Ford, Ph.D.

REFERENCES

Ackerman, N.W. The Psychodynamics of Family Life. Basic Books, New York (1958).

Asberg, M., Cronholm, B., Sjoqvist, F., and Tuck, D. Relationship between plasma level and therapeutic effect of nortriptyline. Brit. Med. J. 3, 331-334 (1971).

Barragan, M. The child centered family, in Family Therapy: Theory and Practice, P.J. Guerin, ed. Gardner Press, New York (1976).

Beck, H. Neuropsychological servosystems, consciousness, and the problem of embodiment. Behav. Sci. 21, 139-160 (1976).

Bell, N.W. and Vogel, E.F. The emotionally disturbed child as the family scapegoat, in *The Family*, N.W. Bell and E.F. Vogel, eds. Free Press, Glencoe, Ill. (1960).

Bertalanffy, L. *General Systems Theory*. Braziller, New York (1967).

Besser, G.M., Delitala, G., Grossman, A., et al. Chlorpromazine, haloperidol, metoclopramide and domperidone release prolactin through dopamine antagonism at low concentrations but paradoxically inhibit prolactin release at high concentrations. *Br. J. Pharmacol.* 71, 569-573 (1980).

Bowlby, J. *Attachment: Attachment and Loss, Vol. 1*. Hogarth Press, London (1969).

Bowlby, J. The making and breaking of affectional bonds. I. Aetiology and psychopathology in the light of attachment theory. *Brit. J. Psychiat.* 130, 201-215 (1977).

Bowen, M. *Family Therapy in Clinical Practice*. Jason Aronson, New York (1978).

Bowen, M., Dysonger, R.H., and Brody, W.M. Study and treatment of five hospitalized family groups each with a psychotic member. *Amer. Orthopsych. Assoc. Proc.* (March 1961).

Carroll, B.J. Clinical application of neuroendocrine research in depression. *Handbook of Biological Psychiatry, Part 3: Brain Mechanisms and Abnormal Behavior - Genetics and Neuroendocrinology*. Marcel Dekker, New York (1980), pp. 179-184.

Chiodo, L.A. and Antelman, S.M. Repeated tricyclics induce a progressive dopamine auto-receptor subsensitivity independent of daily drug treatment. *Nature* 287, 451-454 (1980).

Creese, I. and Sibley, D.R. Receptor adaptations to centrally-acting drugs. *Ann. Rev. Pharmacol. Toxicol.* 21, 357-391 (1981).

Delgado, J.M.R., Grau, C., Delgado-Garcia, J.M., et al. Effects of diazepam related to social hierarchy in rhesus monkeys. *Neuropharmacology* 15, 409-414 (1976).

Fogarty, T.F. Triangles, in *The Family*. The Center for Family Learning, New York (1978).

Gillespie, D.D., Manier, D.H., and Sulser, F. Electroconvulsive treatment: rapid subsensitivity of the norepinephrine receptor-coupled adenylate cyclase system in brain linked to down regulation of beta-adrenergic receptors. Comm. Psychopharmacol. 3, 191-195 (1979).

Gilman, A.G., Goodman, L.S., and Gilman, A. The Pharmacological Basis of Therapeutics. The Macmillan Co., London (1980).

Goodford, P.J. The hemoglobin molecule: is it a useful model for a drug receptor? Trends in Pharmacological Sciences 1, 307-315 (1980).

Haber, S., Barchas, P.R., and Barchas, J.D. Effects of amphetamine on social behaviors of rhesus macaques: An animal model of paranoia, in Animal Models in Psychiatry and Neurology, I. Hanin and E. Usdin, eds. Pergamon Press, New York (1977).

Heard, D.J. From object relations to attachment theory: A basis for family therapy. Brit. J. Med. Psychol. 51, 67-76 (1978).

Henker, B. and Whalen, C. Hyperactive Children. Academic Press, New York (1980).

Jackson, D.D. and Weakland, J.H. Conjoint family therapy, some considerations on theory, technique and results. Psychiatry 2, 30-45 (1961).

Jackson, D.D. Family interaction, family homeostasis and some implications for conjoint family psychotherapy, in Therapy, Communication and Change, D. Jackson, ed. Science and Behavior Books, Palo Alto, Calif. (1968).

Kantor, D. and Lehr, W. Inside the Family. Jossey-Bass, San Francisco (1975).

Klein, D.F. Anxiety reconceptualized. Comprehen. Psychiat. 21, 411-427 (1980).

Liebowitz, M.R. and Klein, D.F. Hysteroid dysphoria. Affective Disorders: Special Clinical Forms. Psychiatric Clinics of North America 2, 555-576 (1979).

Magliozzi, J.R., Hollister, L.E., Arnold, K.V., and Earle, G.M. The relationship of serum haloperidol levels to clinical response in schizophrenic patients. Amer. J. Psychiat. 138, 365-367 (1981).

Miller, J.G. Living Systems. McGraw-Hill, New York (1978).

Post, R.M., Lockfield, A., Squillace, K.M., et al. Drug-environment interaction: Context dependency of cocaine-induced behavioral sensitization. Life Sciences 28, 755-760 (1981).

Pribram, K.H. Feelings as monitors, in Feelngs and Emotions, M. Arnold, ed. Academic Press, New York (1970).

Raleigh, M.J. and McGuire, M.T. Biosocialpharmacology. J. McLean Hospital 2, 73-85 (1980).

Sermons, A.L., Rose, F.H., and Walker, C.A. Twenty-four hour toxic rhythms of sedative-hypnotic drugs in mice. Arch. Toxicol. 45, 9-14 (1980).

Stachowia, K.H. Psychological disturbances in children as related to disturbances in family interaction. J. Marriage and Family 30, 123-127 (1968).

Sulser, F. New perspectives on the mode of action of antidepressant drugs. Trends in Pharmacological Sciences 1, 92-94 (1979).

Winnicott, D. Collected Papers Through Pediatrics to Psychoanalysis. Tavistock Publications, London (1958).

Chapter 13
A Model For Studying Drug Use, and Effects in Dyadic Interactions

M.T. McGuire

INTRODUCTION

This paper presents a model for studying drug uses and effects in dyadic interactions. The model is rooted in evolutionary biology and attempts to integrate selected features of evolutionary theory with clinical findings. Experimental support of the model comes primarily from studies of nonhuman species. The paper begins with a brief discussion of current models which relate to the topic of drug uses and effects in a social context.

It is perhaps surprising that one of the major problems in that subfield of pharmacology which investigates drug effects in social settings -- sociopharmacology -- is that there are too many incomplete models (McGuire et al., 1982). This situation is understandable: sociopharmacology is unusually broad in scope; many questions still need to be asked; and many areas await empirical and theoretical verification and integration. Consider, for example, an experiment in which a drug is given to an animal when alone and, alternatively, when the animal is in a stable social group. Suppose that the frequency of a behavior measureable in both conditions (e.g., locomotion) differs significantly across the two conditions. Suppose further that by systematically varying the age and sex composition of the social group additional significant differences in locomotion are observed. What began as a rather straight forward two-condition study of locomotion in response to a drug yields findings which raise a new set of questions. How should different group compositions be characterized? Are subject animals physiologically altered as a function of different group compositions? If present, can physiological differences explain the observed behavioral frequency differences? What information and/or behaviors result in the postulated physiological differences? Are many or only selected behaviors affected across different study conditions? The questions continue. An investigator would soon find that to make sense of the initial data a large number of additional studies would be necessary. Eventually, followup studies might digress significantly from the initial research question. Indeed, should all the new questions be pursued, a decade might pass before the initial research question could be answered. Because few investigators have the time, the inclination, or the resources to investigate the majority of new questions raised by each of their studies, findings frequently remain tentative and suggestive rather than definitive and models built on such findings are incomplete.

One can argue that science proceeds in this manner. Initial studies, while answering one or two questions, usually raise new questions, some of which are pursued, others not. Yet often this process is haphazard and critical questions frequently remain unaddressed for years. A well constructed model, that is one which integrates theory and data, has a definite place in this environment because it facilitates the development and testing of critical hypotheses. But it is necessary to have a model to start with.

This paper presents the results of an attempt to develop a model which may be used to study drug uses and drug effects in dyadic interactions which meet the following conditions: at least one individual (a) is suffering from atypical psychological and/or physiological-biochemical state(s) and (b) is the recipient of undesirable social exchanges. The points developed in the model could be expanded to include larger numbers of persons. Thus the model could serve as a basis for investigating drug uses and drug effects in groups of three or more people, but this expansion is not undertaken here because to do so would extend this paper significantly. In its present form the model is applicable to people who suffer from psychiatric illnesses generally referred to as neuroses and not to persons suffering from psychoses. A number of variables are critical to the model, including the symptoms of anxiety and depression (these are the only two symptoms that are discussed), psychiatric signs, atypical psychological and physiological-biochemical states, behavioral changes, and social consequences. For this paper it is assumed that these variables have neither strong genetic nor organic (permanent or semi-permanent anatomical change) contributions.

BACKGROUND

In this section I will discuss selected variables associated with drug uses and effects. This discussion will serve to identify the kinds of between-variable relationships which the model has been developed to explain.

Most current pharmacological research in psychiatry focuses on questions relating to the effects of drugs on (a) atypical cognitive states, (b) disturbing symptoms, and (c) behavioral signs (relatively discrete atypical behaviors given a person's age and sex and/or his/her social context). Research is considerably less focused on how drugs affect more general behaviors (e.g., how people make and keep friends) or the social consequences (e.g., loss of friends) which frequently are associated with psychiatric illnesses (McGuire and Essock-Vitale, 1981). The focus of current research is understandable if one assumes the following relationships: atypical psychological and/or physiological states manifest as symptoms and signs and result in behavioral changes which, in turn, often result in social consequences. Given these relationships, it is both reasonable and efficient to design studies to identify drug effects on psychological and/or physiological states, for if these states can be altered, symptoms and signs, general behaviors and social consequences also may be altered.

There are, however, reasons to believe that the relationships described above comprise only a few of many relationships which determine drug uses and effects. These other relationships -- I shall refer to all relationships as "routes" and number each route -- require evaluation.

324   McGUIRE

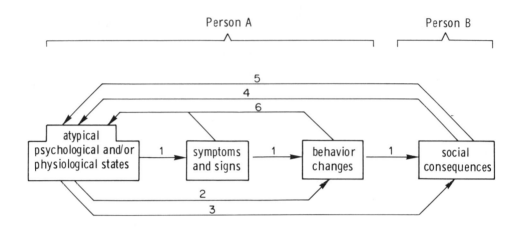

Figure 1. <u>Postulated relationships between variables associated with drug uses and effects in dyadic interactions</u>. This figure depicts six different relationships between variables considered to be critical to understanding drug uses and effects in dyadic (two person) interactions. At least one route is applicable to any interaction but usually several apply. The importance of routes may change over the course of an interaction as described in the text.

Figure 1 depicts six routes between two or more variables which are postulated to influence drug uses and effects. A route may or may not have causal implications. However, the direction of an arrow indicates temporal relationships insofar as they are known. For this paper, the following operational definitions will be used. The term "atypical psychological state" refers to abnormal thoughts (e.g., unusual worries, excessive fears, obsessions) given a person's age, sex, and/or social context. The term "atypical physiological-biochemical state" refers to abnormal physiological-biochemical profiles given a person's age and sex. "Symptoms" are experienced unpleasant affects. While they often are associated with atypical psychological states, this is not uniformly true. However, symptoms are assumed to be associated with atypical physiological-biochemical states. "Signs" are atypical yet fairly discrete behaviors, or the absence thereof (e.g., footshake, unusual posture, lack of facial expression), generally associated with psychiatric disorders. Signs usually are observed by others, not by the person manifesting the sign. For this paper, signs also refer to the failure to "cluster" (synchronized in time and space) behaviors that normally appear together in social interactions, such as looking at one's interaction partner, orienting one's body towards one's partner, using characteristic listening and speaking postures, etc. (McGuire and Polsky, in press). Signs may or may not accompany symptoms. "General behavior changes" refer to acts such as avoiding others, excessively seeking the company of others, and/or making unusual demands. Behavior changes may occur with or without the presence of signs and symptoms or atypical psychological states but atypical physiological states are assumed to be present. "Social consequences" refer to the behaviors of one's dyadic partner which are different than one desires -- generally a reduction in the probability that one's partner will further interact.

Route 1 (Figure 1) has been described above and depicts relationships between physiological and psychological states, signs and symptoms, behavioral changes, and social consequences. Route 2 depicts a situation in which there are physiological state and general behavioral changes. Psychological state changes may occur even though signs or symptoms are not present. A decreasing interest in others accompanied by a decreased frequency of initiated social contact is an example of the events associated with Route 2. So also is the situation in which one unilaterally raises one's expectations of another and acts as though the other should behave according to the expectations. Route

3 depicts a more subtle relationship in which atypical physiological changes occur (psychological changes may or may not occur) in the absence of signs or symptoms and behavioral changes but with social consequences. For example, the meaning of what one says may be frightening to a listener who, in turn, withdraws socially.

The behavior of one's dyadic partner may alter one's psychological and/or physiological states as depicted in Route 4. Route 4 is limited to situations in which the behavior of a partner represents a response to an ongoing or recently completed dyadic interaction. Events associated with Routes 1 or 2 or 3 may then occur. Route 5 depicts a situation in which the behavior of a partner also affects one's psychological and/or physiological states but the partner's behavior is not primarily in response to an ongoing interaction, i.e., one's partner behaves in unexpected and/or unreasonable ways, as in situations of "displaced anger." Route 6 depicts the effects of one's behavioral changes and/or symptoms on one's psychological and/or physiological states. One often becomes depressed over being anxious or socially ineffectual just as one may become anxious during periods of depression if one feels compelled to engage in behaviors which one would rather avoid. There is, of course, every reason to believe that more than one route in Figure 1. is operative over the course of any extended dyadic interaction. Moreover, different combinations of routes may characterize relationships with different partners.

While the focus of this model is on humans, experimental evidence supporting the routes postulated in Figure 1 is most readily available from studies of nonhuman species. This evidence has recently been reviewed (McGuire et al., 1982), and only salient points will be discussed here. In nonhuman primates, the frequency of a large number of social behaviors can be altered by giving drugs to either normal animals or animals who have been raised in social isolation (depressed) (Routes 1 and/or 2). Alteration of an animal's behavior through drugs alters the behavior of nontreated animals in the treated animal's group towards the drugged animal (Routes 1 and/or 2 and/or 3 and 4). Alterations of the behavior of animals through drugs alters the physiological and biochemical profiles of non-treated subject animals (Routes 4 and/or 5). Alteration of the behavior of animals through altering their status (and thus behavior) results in changes in biochemical profiles (Route 6), and prevention of an animal (via social manipulation) from engaging in certain behaviors has striking effects on biochemical and physiological profiles (Route 6).

It is important to emphasize that dyadic interactions take place within a context of established, sometimes shared, expectations. For example, in the conventional greeting, "Hi, how are you?", it is expected that the question will not be taken too literally. Only a short reply is in order. Even in long established relationships where behavioral expectations are relationship-specific, expectations still prevail. In general, it is the violation of expectations that results in social consequences. (Reasons why this is so are further discussed below.) Another point requiring emphasis is the link between the behavior of one's dyadic partner and one's physiological and/or psychological states (Route 4, Figure 1.). Behaviors by others suggesting approval, empathy, rejection, and/or indifference have significant impacts on all of us and these impacts often continue long after interactions have ended and thus may affect subsequent interactions.

If one accepts that the routes shown in Figure 1 (or a similar representation) depict the basic kinds of temporal and feedback relationships which occur between selected variables in dyadic interactions, then it is perhaps clear why dyadic interactions might be the most appropriate experimental unit on which to conduct detailed research on drug uses and effects in social settings. To study drug uses and effects in groups of three or more individuals would introduce considerably greater complexity, and systematic studies may be unusually difficult given our present state of knowledge. This point is illustrated in Figure 2 which shows findings from studies by Fairbanks et al. (1977) and

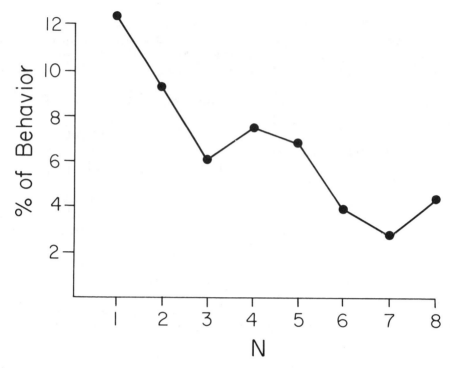

Figure 2. <u>Changes in the frequency of psychiatric signs per individual patient as a function of the number of people in a given area.</u> This figure was developed through random focal sampling techniques and represents findings from more than 12,000 samples over an 8 month period. The effect is independent of room size and of activities carried out in the room.

McGuire et al. (1977) in which the frequency of signs for psychiatric inpatients was measured as a function of the number of people in a given hospital area (e.g., recreation, relaxation, eating areas).

In Figure 2 the frequency of signs per subject--all subjects were receiving psychotropic medications--shows a general downward decline as the number of people in a given area increases. These changes occurred even when subjects were passively socializing, as when one is temporarily a silent member of a multi-person group. The frequency of specific signs declines even further, while others increase, when patients are involved in dyadic interactions (Polsky and McGuire, 1981). These findings not only point to the critical importance of knowing and/or setting specific conditions for studies of drug effects when social behavior is a dependent variable, but they also serve as a possible warning to investigators assessing drug effects in social contexts: namely, a drug may appear to have quite different effects depending on the research context.

BASIC ASSUMPTIONS OF THE MODEL

In this section the basic assumptions of the model are reviewed. The most basic assumption of the model is that man is predisposed to optimize biological goals. The evidence and ideas supporting this view have been reviewed in detail elsewhere (Alexander, 1980; McGuire and Essock-Vitale, 1981), and may be briefly summarized as follows: While the most fundamental biological goal is that of genetic replication (see Wilson, 1975 for a review), a number of related goals (e.g., developing social support systems, living in an optimally dense environment, protecting oneself from predators, staying healthy, etc. -- see Appendix A for a full listing), are essential to the multi-faceted and complex process of replication. If related goals are optimized, the probability of genetic replication increases. Thus, it is often more important to assist others in order to establish reciprocal debts, have others provide emotional support, or to acquire material resources than it is to engage directly in replication (e.g., having offspring, investing in offspring and/or investing in genetic kin). <u>Optimizing nearly all biological goals requires some form of interaction with one or more persons, and frequently goals are optimized via dyadic relationships</u>. Hence the critical importance of social relationships in evolutionary theory and the equally critical importance of reciprocity in extended relationships.

A second assumption is that alterations in psychological and physiological states and associated symptoms (depression and anxiety) result for one of two reasons: either because one or more biological goals <u>have not</u>, <u>are not</u> -- the basic reason for <u>depression</u> -- or because, in the anticipation of the actor, one or more goals <u>will not</u> be optimized -- the basic reason for <u>anxiety</u>. Consciousness of these relationships is not implied. The symptoms associated with depression thus reflect past, ongoing, and anticipated suboptimal dyadic interactions while the symptoms associated with anxiety reflect the anticipation of one or more suboptimal dyadic interactions. One can argue that the reasons for depression and anxiety and their associated psychological and physiological states are considerably more diverse than those being suggested here. The evidence from the in depth study of individuals (psychoanalysis), seems convincing on this point, however: benefits and costs are nearly always associated with other people although they may initially manifest in consciousness in forms which suggest a dissociation from others. (There are a number of other posited explanations for these symptoms as well as associated psychological and/or physiological changes, such as those proposed by psychoanalysts, learning theorists, and labeling theorists. These explanations are not necessarily incompatible with the view presented here -- see especially Alexander, 1980; and McGuire et al., 1981).

Biological goals are seldom optimized with a single behavior or even a specific set of behaviors. Rather, optimization is achieved in incremental steps by varying behavioral choice, sequence, and intensity as a function of one's motivation and social and environmental options. Thus, alterations in psychological and physiological states, anxiety, and depression are most appropriately viewed as responses to not having achieved, not currently achieving, or anticipating not achieving one or more incremental steps.

The third assumption is that the decision to engage in a behavior can be viewed in benefit/cost (B/C) terms. Operationally, optimization can be said to occur whenever the B/C ratio exceeds 1 (B/C-1) for achieving a given incremental step.

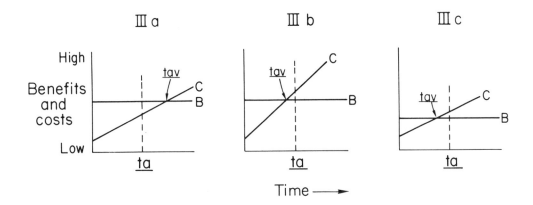

Figure 3. Benefit/cost curves associated with specific behaviors. This figure shows three postulated benefit/cost curves representing anticipated costs and benefits for normal persons (3a) and for persons who suffer from atypical psychological and/or physiological states (3b-c). Details of the curves are explained in text.

Figures 3a-c show idealized representations of B/C curves and their postulated relationships to physiological and psychological states, symptoms, and behaviors.

Figure 3a depicts a "normal" B/C curve for carrying out a specific behavior associated with an incremental step in a dyadic interaction. B represents the anticipated benefit from the behavior, i.e., achieving an incremental step such as obtaining an important piece of information from one's dyadic partner. C represents the anticipated accumulated costs over time from engaging in the behavior. t_a represents the anticipated average amount of time required to obtain the predicted benefit given one's prior experiences. t_{av} represents the anticipated maximum amount of time available to complete a behavior and gain the benefit while the anticipated benefit exceeds the cost (B/C-1). Note that in Figure 3a, t_{av} is to the right of t_a, a situation in which a behavior is likely to be enacted. There is extra time ($t_{av}-t_a$) beyond the average (t_a) to achieve the incremental step with a B/C-1. Because of differences in experience, the motivation to achieve an incremental step, age, sex, and behavioral capacities, t_{av}, t_a, C, and B all will differ from person to person and, for a given person, across dyadic partners.

In Figure 3a, benefits are shown as remaining constant over time while costs accumulate at a constant rate. However, in actual interactions, B and C probably change from moment to moment as a function of the characteristics of the interaction. Direct costs would be determined by the summation of physiological and psychological requirements for engaging in the behavior as well as the costs of inhibiting other behaviors in which one might engage and which might be beneficial. Indirect costs would be measured in lost opportunities to enact other potentially beneficial behaviors and possible adverse consequences resulting from engaging or not engaging in the behavior in question.

A nearly infinite number of variations of Figure 3a can be imagined. For many reasons costs and benefits may be greater or less than anticipated. Which possibilities will apply will be determined in part by one's capacity to anticipate accurately the behavior of one's dyadic partner. Thus, in an X-Y interaction, X may desire a piece of critical information (B, Figure 3a) from Y; X also may anticipate that twenty minutes of conversation with cost C will be required to gain the information (t_a Figure 3a). Suppose, however, that Y does not have the information, and X spends 15 minutes determining this fact. The benefit to X would be 0 and costs would be considerable. Or, it may take X 40 minutes (assume t_{av} is 30 minutes) to obtain the information, in which case the costs may exceed benefits. Other things being equal and over an extended period of time, X should increasingly engage in behaviors associated with a high B/C, seek dyadic partners in which the sum of B/C ratios is likely to be >1, discontinue behaviors which repeatedly have a B/C <1, discontinue relationships in which the sum of B/C ratios is likely to be <1, and seek new strategies for achieving incremental steps when the B/C approximates 1.

Figure 3b is one of two ways of developing a B/C characterization of changes in psychological and/or physiological states, symptoms, and behaviors. As the figure is drawn, benefits remain the same as in Figure 3a, but costs accumulate at a more rapid rate than usual, resulting in a reduction in the time available (tav) to achieve a benefit with a B/C >1. Other things being equal, the postulated consequence of the left shift of tav is a reduced probability of the occurrence of the behavior associated with the 3b configuration. Why would a person anticipate that costs will increase? In principle, this could occur because (a) of a misinterpretation of the costs required to achieve a benefit, (b) because one cannot efficiently enact the required behavior to achieve a benefit (lacking a behavioral skill), (c) being in a situation in which the actual costs are higher than usual (a stubborn dyadic partner), and/or (d) because of costs required to inhibit competing and otherwise desirable behaviors. Depression can be associated with Figure 3b for any subset of the reasons mentioned above but a cost debt (the summation of recent dyadic interactions in which the B/C <1) also must be present. This debt is postulated to affect one's psychological and/or physiological states as well as associated B/C calculations. For example, persons who are depressed often appear to have normal benefit expectations yet they are unlikely to engage in a behavior because the existing cost debt is already substantial. Anxiety also can be associated with Figure 3b. It is postulated to be the result of anticipating that a behavior which one feels compelled to engage in will result in a B/C <1. No existing cost debt is postulated for anxiety.

However, cost debts may occur in persons who primarily are anxious if they repeatedly fail to act, a point which may in part explain the frequent association of symptoms of anxiety and depression.

Figure 3a represents a situation in which anticipated benefits are reduced while the costs for enacting a behavior remain the same as in Figure 3a. As in Figure 3b, $\underline{tav}$ is to the left of $\underline{ta}$, thus reducing the probability that a behavior associated with Figure 3c will be enacted. There are a number of reasons for the low benefit line. When associated with anxiety, the primary reason appears to be that one feels that one will not efficiently or appropriately engage in a behavior, thus reducing the probability of one's dyadic partner acting in a desired way. When associated with depression, prior experiences in which B/C-1 result in anticipated low benefits. Further, an already depressed person may avoid accepting benefits (e.g., assistance) because the acceptance may obligate him/her to reciprocate at a later time. (A lowered benefit line also may represent a realistic assessment of events, a point especially applicable to long standing dyadic relationships, such as marriage, where one partner no longer satisfies the other.) Combinations of Figure 3b and 3c are more probable than not, and particularly for certain types of depression where anticipated benefits are reduced and costs are increased. Possible combinations will not be discussed in detail here.

For Figures 3a-c errors in anticipated costs and benefits can occur but this is probably a minor issue with respect to persons suffering from neuroses. However, one consequence of severe psychiatric illness (e.g., psychosis) appears to be consistent B/C calculation errors in one or another direction. In hypomania, for example, the benefit line appears to be significantly and unrealistically elevated. This elevation may explain why persons with this illness initiate so many behaviors (Polsky and McGuire, 1979). Yet with such persons either the B line rapidly drops or the C line rapidly increases once a behavior is initiated, thereby resulting in the short duration of behaviors (Polsky and McGuire, 1979).

	Figure 3a "normal"	Figure 3b depression	Figure 3b anxiety	Figure 3c depression	Figure 3C anxiety
Predisposition to behave	1	1	1	1	1
Postulated cost debt	0	+	0	+	0
Cost of inhibiting the behavior to which one is predisposed	.25	.25	.25	.25	.25
Anticipated benefit if the behavior is enacted	1	1	1	.75	.75
Anticipated cost if the behavior is enacted	.50	.90	.90	.50	.50
Time required to achieve the benefit under normal conditions (t_a)	1	1	1	1	1
Time available to achieve a benefit under different conditions (t_{av})	1.25	<1	<1	<1	<1
Probable B/C if behavior enacted	>1	<1	<1	<1	<1

Table 1. <u>Summary of Figures 3a-c.</u> This table summarized Figures 3a-c in B/C terms. (*) = cost debt must be added to probable B/C. Details are discussed in text.

Table 1 summarizes postulated differences between depression and anxiety and Figures 3b-c. Recall, however, that symptoms need not occur with Figures 3b-c for Routes 2 and 3 (Figure 1). Thus, Table 1 is only partially applicable to Figure 1. Throughout Table 1 it is assumed that the predispostion to engage in a behavior remains constant across all conditions. In reality this assumption is unlikely to hold because of differences in motivational states, the importance of an incremental step, environmental options, past relationships with one's dyadic partner, etc.

As shown in Table 1 for Figure 3b, anxiety is associated with an anticipated B/C-1, and no cost debt. The anxious person is thus faced with a dilemma: to act as he/she feels compelled to do will not result in a B/C-1 but not to act will result in costs for inhibiting the predisposition to act and the indirect costs for not acting. Depression is associated with the Figure 3b configuration when there is an already existing cost debt. The depressed person is faced with another type of dilemma: to act will not result in a B/C-1 but not to act will further raise the already existing cost debt. For Figure 3c the situation is essentially similar to Figure 3b. As depicted, however, the probability of acting is reduced compared to Figure 3b because of the reduction in anticipated benefits. This formulation may explain why depressed individuals often withdraw from social contact: to do so reduces stimuli, which, if responded to, would result in a B/C-1. It may also explain why anxious persons are more likely to act than depressed persons: their decision is not complicated by cost debts.

To summarize thus far, for both Figures 3b and 3c the probability that a behavior will occur is reduced primarily because the anticipated time available ($\underline{t_{av}}$) to obtain benefits from the behavior is to the left of $\underline{t_a}$. This situation is not necessarily eased by selecting less appropriate behaviors (e.g., less efficient behaviors given a particular incremental step) because attempts to achieve incremental steps by indirect routes may increase costs.

Do persons consciously or unconsciously engage in B/C calculations, as is suggested in Figures 3a-c? There is, of course, no fully satisfactory answer to this question. One can engage in such calculations consciously, but consciousness of the points under discussion is neither implied nor required in this model. The assumption that such calculations (or some equivalent) are made is not inconsistent with current observations however. A possible insight into this question is provided by the work of Plutchik and his colleagues (1962, 1980) who discuss emotions -- they use the word in a broad sense -- in terms of their adaptive functions as well as their <u>importance</u> to individuals, not their symptom-related characteristics. In their view, when emotions such as gregariousness and trust (emotions commonly associated with dyadic interactions) are reduced in importance, persons are less likely to enact behaviors normally associated with these emotions (e.g., initiate discussions, respond to others, etc.). Evidence supporting this view is now available (McGuire and Polsky, unpublished data) and the findings may be taken to indicate the consequences of B/C calculations which are <1. A clear implication of Plutchik's work is that an emotion often generalizes to situations not experientially related to the emotion, e.g., one withdraws from friends where the B/C ratio consistently has been <1. This implication is consistent with clinical experience with severely ill individuals: transient depression and/or anxiety are often limited to a single dyadic partner or event but more pervasive symptoms tend to generalize. However, the type of person being discussed in this model (people suffering from neuroses) generally make clear distinctions between dyadic partners and they will enact some behaviors with some partners but not others. Thus, a defensible postulate to apply to Figure 1 is that the B/C calculations for a given behavior should change in different ways as a function of one's cost debt, the behavior of self, the behavior of one's dyadic partner, and the importance of an incremental step with which the behavior is associated.

The fourth assumption of the model is that drugs are used by persons to alter B/C outcomes. On first examination, this postulate seems unlikely, because many (if not most) persons who use drugs will claim that they do so primarily to relieve symptoms. As mentioned earlier, detailed interrogation of people suffering from the symptoms under discussion here suggests that these symptoms are associated predominantly with social interactions. Statements such as "I don't want to see anyone," "I am so anxious I can't relate," and "I get nervous everytime I talk with him" are examples. Likewise for behavioral changes: statements such as "I don't feel like working," "I never go to parties any more," and "I can't seem to get organized" all suggest behavioral changes which have social implications in that these behaviors may displease others. Moreover, it is symptoms (if present) and/or behaviors which people wish to alter. They sense that behavioral changes and symptoms adversely affect their behavior and social relationships. Operationally, therefore, persons in the population being discussed may be assumed to take drugs in order to alter B/C calculations which, in turn, alter one or more of the following: psychological and physiological states, symptoms, behaviors, or social consequences. To argue in this fashion is not to deny the affective component of symptoms (when symptoms are present). Rather, the point is to emphasize that unpleasant affects and behavioral changes nearly always have an undesirable social component. Often during periods of depression thoughts are of past events. But it is unlikely that the population under discussion takes medications because of unpleasant past events, but because of the cost debt associated with such events. If one talks with depressed persons they will eventually reveal that they view most social interactions as having a B/C-1. Moreover, as a group, depressed persons have limited social-support networks (e.g., essentially, friends or kin from whom they can receive support -- Brown et al., 1975; Henderson, 1977). This situation reduces their chances of engaging in possible high-benefit dyadic interactions. While similar findings apparently are not available for persons suffering from anxiety, clinical evidence suggests that a similar type of social isolation exists for many such persons.

THE BASIC ELEMENTS OF THE MODEL

I will now turn to the details of the model by way of describing an evolutionary biological view of dyadic interactions. Following this description I will suggest how the B/C analysis of interactions may be helpful in devising research designed to elucidate drug uses and effects.

Consider a situation in which two adults, X and Y, interact with one another. MacKay (1972) has characterized such situations in the following way. X will interact with Y for one of two reasons: (a) either to confirm what X already knows or believes (e.g., that Y is a friend, that Y will still repay X an owed favor, etc.) or (b) to gain something that is advantageous to X (e.g., a piece of information, physical help, emotional support, etc.). Kummer (1978) has further developed this model particularly with regard to the assumption that man is predisposed to optimize biological goals. We shall now examine in some detail what I shall refer to as the MacKay-Kummer model of dyadic interactions. But before doing so it should be noted that a somewhat similar model has been developed by Emerson (1962, 1972) and Cook and Emerson (1978) and is often referred to as power-dependency theory. The Emerson-Cook model builds on the idea that in a dyadic interaction A has control (power) over B to the degree that A controls the outcome of the interaction. Or, A's power over B is equal to B's dependence on A for reinforcer Z, and B's power over A is equal to A's dependence on reinforcer W (see also Molm, 1981; Lamb, 1981). In power-dependency theory, power is a structural characteristic of a dyadic interaction, whereas power use is a behavioral characteristic of the actors. While much of what is said below can be stated in power-dependency terms, I have elected to discuss dyadic interactions from an evolutionary perspective. Evolutionary theory both addresses a broader number of theoretical variables and a more extensive data base than power-dependency theory, including the social consequences of behavior and the conditions which may affect interaction outcome.

The first and perhaps the most obvious implication of the MacKay-Kummer model is that X is dependent on Y to enact certain behaviors which are essential to X optimizing X's incremental steps. In all probability, a second implication is as follows: the more specific X's requirements for Y's behavior, the less likely X is to gain exactly the desired benefit from Y. This would follow on the basis that Y also is seeking benefits in an interaction with X and does not engage in an interaction simply to behave in a manner suitable to X. X thus must be able to assess the probability of Y behaving in ways X desires and in which the B/C-1. In dyads involving unfamiliar persons, X may spend considerable time and effort (costs) assessing the probability of Y behaving in ways advantageous to X. This point may explain why people are often wary of entering new relationships -- an initial cost debt may develop in the process of determining if a person is a potentially worthwhile partner. Assessment costs should be reduced in familiar relationships, however. There are, of course, many important variations on these points. For example, children often do not bother to assess their mothers' willingness to provide favors before requesting them. And, one must be cautious in asking even a close friend to grant a favor if one already is in debt to the friend. In evolutionary terms, it is possible to make specific predictions about the probability of Y behaving as X desires as a function of X and Y's genetic relatedness (Hamilton, 1964). These predictions will not be discussed here.

The third, and probably the most important implication of the MacKay-Kummer model is as follows: there are reciprocal effects of X on Y and Y on X. The bidirectional components of dyadic interactions cannot be divorced easily. Y obviously has considerable "control" over whether X achieves a desired incremental step with a B/C-1. But if X wants something and Y can determine whether or not to give it, X's control of the interaction is still not 0 because Y may wish X to reciprocate if Y behaves as X desires. (That debts may be owed for granting favors is not always made explicit in interactions, but in all probability the assumption of reciprocity is present.) Assuming that X has initiated an interaction, Y's behavior will in varying degrees be a function of how X behaves. For example, X may offer future favors before asking Y to behave in a certain way. To do so may have the effect of communicating the importance X places on Y's behavior.

A fourth implication is that errors are possible. For example, X may ask and gain a favor from Y on the condition (set up by Y or because of the structure of the interaction) that X is willing to reciprocate. X may assume that the required reciprocation is either greater or lesser than Y assumes; or, X may not recognize that repayment is expected, in which case there may be social consequences. A related implication--the fifth-concerns deception. Viewing interactions from the perspective of individual selection, it is to X's advantage to minimize the cost of an interaction while maximizing benefits. But X should also try to minimize possible future costs (owed debts) as well as possible consequences of not reciprocating. This situation invites deception or the suggestion by X to Y that a favor either will be or is being repaid when full repayment is not intended. Thus, Y must assess the likelihood of X's reciprocation before behaving in the way X desires. Overt behavior may be an important variable with regard to this point because, as a number of investigators have shown (e.g., Ekman et al., 1976), certain nonverbal behaviors often are interpreted as indicators of deceptive intent.

The preceding points appear to apply to persons suffering from psychiatric illnesses. Generally, such persons behave atypically. This is quite obvious in cases of extreme illnesses, such as the psychoses, but it also is true among people suffering from anxiety or depression: behaviors may be enacted suboptimally (McGuire and Essock-Vitale, 1982), voice quality may be atypical, the rate of speech may be increased or decreased, and interactions may be excessively intense, demanding, or unresponsive. Such behaviors represent deviations from expected interaction conventions which generally are essential to successful interactions (e.g., B/C-1). Certain behavioral conventions, such as a head-nod, a smile, or a gesture indicating Y's agreement with X suggest that benefits are forthcoming; conversely, other conventions, such as excessive questioning, unresponsiveness, looking elsewhere, not picking up a point in the conversation, etc., suggest that benefits will not be forthcoming. Deviations from the expected are potentially costly because they usually suggest that benefits will not be forthcoming. At times they are interpreted as indications of deception. Individuals engaging in deviations thus represent "bad bets" -- B/C-1 -- for social interactions. It follows that they will be avoided (see McGuire et al., 1982).

Consider further the following situation: assume that X has assessed that Y is unlikely to behave as X desires and that X wishes to alter Y's behavior. X's problem becomes one of altering Y's B/C estimates with respect to X's desired behavior. X may try to do so through persuading, promising, confronting, arguing, making guilt inducing statements, posturing, threatening, gesturing, etc. (Trivers, 1971; Kummer, 1978). To determine if Y has been altered, X must assess the effects of X's behavior on Y. To do this, X not only requires an accurate estimate of Y's characteristic behaviors but also a way of estimating how rapidly Y is changing. As MacKay suggests, it seems likely that we also infer others' motives from their behaviors and that we often behave in ways to confirm or disconfirm the suspected motives of others. Frequently, therefore, interactions may get diverted to attend to these matters. Needless to say, an elaborate ongoing B/C computational system is essential.

The preceding points can be restated in B/C terms. If Y behaves as X anticipates, X should continue the interaction until the benefit is received or until it is no longer beneficial to continue an interaction because X calculates that he/she can optimize benefits elsewhere. If the benefits of the interaction exceed X's expectations while X's costs remain as initially anticipated, X should extend the interaction, perhaps also explore new ways of exploiting it. Conversely, if the benefits appear to be less than X anticipated, X may either interrupt the interaction or develop new strategies with respect to Y. The probability of developing new behavioral strategies should be a function of the importance of Y's behavior to X. Because the preceding points apply to Y as well, both partners are assumed to monitor an interaction, all the while comparing the actual B/C to the anticipated B/C and modifying their behavior as new information appears. Their strategies thus are interdependent; the strategy which X adopts will be influenced by Y's strategy and vice versa (see Maynard Smith, 1971, for an extended discussion on this point).

The possible costs and benefits associated with a dyadic interaction can be summarized as follows. Potential costs for X include: (a) costs for initiating a behavior; (b) costs for inhibiting competitive behaviors; (c) costs of monitoring Y's behaviors; (d) costs associated with trying to change Y's behaviors; (e) costs of maintaining a behavior when costs exceed benefits (to the right of point tav in Figures 3a-c); (f) costs to X if X must change his/her behavior during an interaction with Y; (g) potential costs incurred because of promises extracted from X by Y; and (h) costs incurred if Y deceives X into thinking X has or will receive a benefit that is not forthcoming. Potential benefits to X include the achievement of an incremental step and the development of a dyadic relationship which may have predictable benefits during future interactions.

Returning to Figure 1, Route 4 depicts Y's effect on X which either may increase or decrease X's actual costs or benefits. Route 5 depicts an unanticipated interaction which Y initiates towards X. Such interactions may be either positive or negative and they often have dramatic effects, primarily because X is unprepared for Y's behavior.

Route 3 depicts X's initiated interactions with Y. A benefit to X which costs Y very little, such as a confirmation of friendship through a friendly wave, may be easily obtained, but a significant benefit which would require Y to temporarily change Y's plans might be difficult to obtain and depend on the degree to which Y owes X favors and/or Y wishes to satisfy X and gain a future favor.

Route 2 depicts X's behavioral changes which are obvious to Y but not necessarily fully obvious to X. Such changes occur when X is unexpectedly secretive, affectionate, demanding, or withdrawn, in which case potential benefits may be reduced to X but not necessarily costs, because X may not have available alternative ways to achieve incremental steps. Route 6 represents feedback for Routes 1 and 2.

To summarize, recall that humans are predisposed to act so as to optimize biological goals, that many of these goals must be optimized through social interactions, that reciprocity is integral to such interactions, and that achieving incremental steps can be conceptualized in B/C terms. Given these arguments, we have a basis of explaining why and how people interact. We also have a basis for explaining interaction-related conflicts: on one hand, X is predisposed to socialize and use others to optimize his/her incremental steps; to do so involves certain costs, however. Because of B/C calculations, X may anticipate a B/C-1 in which case X is unlikely to enact a behavior, thereby reducing the possibility of optimizing biological goals.

It is through the use of drugs that one may attempt to resolve such situations, that is, alter B/C calculations and outcomes.

HOW CAN THE MODEL BE USED TO UNDERSTAND DRUG USES AND EFFECTS IN DYADIC INTERACTIONS?

Although there exists a variety of postulated relationships between the variables under discussion, it is still true that individuals, not dyads, take psychotropic medications. In the model being developed here, only physiological or psychological states are directly effected by medications. In turn, Routes 1, 2, 3 (Figure 1) may be affected. Routes 4, 5, are affected only secondarily. In principle, effective drugs should influence all routes. However, influences may take place slowly. For example, if X is greatly in debt to Y, Y may be skeptical of changes which take place in X as a result of medications, thereby slowing Route 5 (Figure 1) changes.

In the model presented here, there can be four immediate consequences of psychotropic medications: to increase or decrease anticipated benefits or to increase or decrease anticipated costs. These changes would result from changes in psychological and/or physiological states. To the degree that a medication is effective -- an increase in B and/or decrease in C (Figure 3) -- point $\underline{tav}$ in Figures 3b and 3c should shift to the right. Thus, the time available to complete a behavior while the $B/C > 1$ is extended. As a result, there should be an increase in the frequency of those behaviors in which previously the $B/C < 1$ (provided they are related to incremental steps). If the medication effects proceed only along Route 1 (Figure 1) there should be decreases in symptoms and signs as well as behavioral changes, these being the result of changes in physiological and/or psychological states. However, Routes 1 and 2 may be operative simultaneously or Route 2 may be operative separately. (That Route 2 operates separately from Route 1 is suggested by the observation that behavioral and social consequence changes often are observed while patients report essentially no relief in symptoms.) This line of reasoning can be pursued for all the various possible route combinations in Figure 1.

If persons are predisposed to optimize biological goals, then the most fundamental reason they should take medications is to enhance optimization, i.e., alter B/C outcomes. Conversely, they should discontinue and/or change medications if the drugs appear to them to impair optimization. Because optimization is not assured with either symptom or behavioral changes -- social consequence changes are required -- symptom change cannot be expected to explain satisfactorily drug uses. Uses (which include taking amounts of prescribed drugs as well as nonprescribed amounts) should be closely associated with B/C calculations and actual benefits and costs during interactions.

A number of qualifications apply, however. First, no medication is specific to a particular physiological system, thus any medication will effect systems in addition to the target system. Depending on the systems affected, new B/C considerations may apply. Second, medications work in different time frames. For example, the use of tryptophan to treat depression is based on the logic of providing a precursor to serotonin. Other antidepressant drugs, such as Elavil, are thought to work by affecting compensatory physiological systems. Thus, the effects of different drugs on B/C calculations should differ. Third, during the stages of the natural history of an illness, different medications may be appropriate because of the changing relationships between various physiological systems (McGuire et al., 1982). Fourth, because B/C ratios associated with psychiatric and nonpsychiatric states differ across individuals, it is to be expected that the same medications will have different effects on any two individuals.

Some related points deserve brief mention. In this model, it is possible to obtain temporary symptom relief through medication use but not necessarily to reduce the basic "cause" of one's physiological and/or psychological states if that "cause" is the behavior of another (Y in the previous discussions). It follows that in differing degrees treatment effects will be a function of the characteristics of Routes 5 and 6 (Figure 1). Second, "anti-depression" medications may be more effective than "anti-anxiety" medications in relieving some anxiety symptoms. This should occur in relationship to Figure 3b when both depression and anxiety are present. Conversely, anti-anxiety medications may reduce depression-related symptoms associated with Figure 3c. Recall however, that combinations of Figures 3b and 3c are more likely than either alone.

IMPLICATIONS OF THE MODEL FOR RESEARCH

There are numerous research implications which follow from this model. Perhaps the most obvious applies to selecting study populations. It would seem essential to develop a means of assessing in detail the kinds of dyadic interactions in which subjects engage. Are they "free" to select dyadic partners? What type of dyadic relationship do they have? Are relationships primarily with kin (if so, what is the genetic relationship) or with nonkin?, etc. It would also seem essential to characterize subject populations in terms of the biological goals which they are attempting to optimize at the time they are being studied. Two women of the same age, one with three offspring and the other with none, will probably be depressed for quite different reasons.

A major strength of the model is the centrality of B/C calculations. An obvious research hurdle concerns the means of assessment of B/C calculations outside of contrived situations. As yet, there is no generally accepted way to measure these calculations. Studying B/C calculations may be simplified initially by setting all costs and benefits as anticipatory and disregarding cost debts. For example, by examining a number of two-choice alternative questions dealing with interactions it should be possible to develop for a given subject a B/C ratio for particular dyadic partners whose ways of interacting are well known to the subject. Such ratios could then be compared with actual events and/or the desired or actual use of medications preceeding particular dyadic interactions. Some preliminary work in this area already has been reported (Stizer et al., 1981; Janowsky et al., 1979). It is also possible that physiological and biochemical measures are possible. The animal evidence summarized earlier strongly suggests this possibility (see McGuire et al., 1981, for a review).

A third point concerns the effects of medications over time. A clear implication of the model is that the B/C characteristics of two or more dyadic relationships will change at different rates even if X changes at a near constant rate in both relationships. Specific dyadic relationships and their association with medication effects thus must be monitored over extended periods in order to assess these differences as well as to determine medication effectiveness. Finally it should be noted that in this model the long-term outcome measure is a reduction of negative social consequences in dyadic interactions. Drugs which reduce negative consequences thus should be prefered over those which do not.

By way of concluding, several points should be noted. Although there are many other variables which relate to the points discussed in the model, such as guilt, self-esteem, and different reinforcement contingencies, their inclusion in the model would require a major extension of the paper. Thus, they will not be discussed except to say that in many instances these variables appear to be secondary and thus should not significantly alter the points made above. Moreover, treatment methods other than drugs designed to alter B/C calculations and outcome should be effective (see, for example, Hall and Goldberg, 1977). As developed, the model does not account for why symptoms appear with some individuals but not with others. One possibility is that when symptoms first appeared during childhood they were adaptive (see Lewis, 1934). With children, mild anxiety or depression often lead to empathetic and supportive responses by sensitive parents. Such responses may serve to "reinforce" symptom use. Among adults, similar responses are often provided by kin and friends, at least for limited periods of time. However, as was noted above, over extended periods interacting with persons with symptoms is generally associated with interactions with a B/C-1.

APPENDIX A

List of Biological Goals: Adapted from McGuire and Essock-Vitale (1981).

1. Live in an optimally dense environment.

2. Have an optimal number of offspring.
3. Establish and maintain pair bonds.
4. Acquire and control resources.
5. Stay healthy.
6. Have adequate defenses from predators.
7. Communicate fluently.
8. Develop and maintain social support networks.
9. Develop behavioral and learning flexibility.
10. Optimize investment in offspring and other kin.

ACKNOWLEDGEMENT

Funding for this project was provided by the Harry Frank Guggenheim Foundation.

REFERENCES

Alexander, R. Darwinism and Human Affairs. University of Washington Press, Seattle (1980).
Brown, G.W., Bhrolchain, M.N., and Harris, T. Social class and psychiatric disturbance among women in an urban population. Sociology 9, 225-234 (1975).
Cook, K.S. and Emerson, R.M. Power, equity and commitment in exchange networks. Amer. Soc. Rev. 35, 203-222 (1978).
Ekman, P., Friesen, M.V., and Scherer, K.R. Body movement and voice pitch in deceptive interaction. Semiotica 16, 23-27 (1976).
Emerson, R.M. Power-dependence relations. Amer. Soc. Rev. 27, 31-40 (1962).
Hall, R. and Goldberg. The role of social anxiety in social interaction difficulties. Brit. J. Psychiat. 131, 610-615 (1977).
Hamilton, W.D. The genetical evolution of social behavior. J. Theor. Biol. 7, 1-16 (1964).
Henderson, S. The social network, support and neurosis: The function of attachment in adult life. Brit. J. Psychiat. 131, 185-191 (1977).

Janowsky, D.S., Cloptin, P.L., Leichner, P.P., Abrams, A.A., Judd, L.L., and Pechnick, R. Interpersonal effects of marijuana. Arch. Gen. Psychiat. 36, 781-785 (1979).

Kummer, H. On the value of social relationships to nonhuman primates: a heuristic scheme. Soc. Sci. Info. 17, 687-705 (1978).

Lamb, T.A. Nonverbal and paraverbal control in dyads and triads: Sex or power differences. Soc. Psychol. Q. 44, 49-53 (1981).

Lewis, A.J. Melancholia. J. Ment. Sci. 80, 277-381 (1934).

MacKay, D.M. Formal analysis of communicative process, in Non-verbal Communication, R.A. Hinde, ed. Cambridge University Press, New York (1972), pp. 3-25.

Maynard Smith, T. Evolution and the theory of games. Amer. Scient. 64, 41-45 (1976).

McGuire, M.T. and Essock-Vitale, S. Psychiatric disorders in the context of evolutionary biology: A functional classification of behavior. J. Nerv. Ment. Dis., 1981.

McGuire, M.T. and Polsky, R. The use of hierarchial cluster analysis in detecting behavioral change in acute psychiatric illness, in Sociopharmacology: Drugs in Social Context, C. Chien, ed. Reidel, Dordrecht, in press.

McGuire, M.T., Essock-Vitale, S., and Polsky, R. Psychiatric disorders in the context of evolutionary biology: An ethological model of behavioral changes associated with psychiatric disorders. J. Nerv. Ment. Dis., 1981.

McGuire, M.T., Raleigh, M.J., and Brammer, G. Annual review of sociopharmacology. Annual Reviews Inc., Palo Alto, in press a.

Molm, L.D. Power use in the dyad: The effects of structure, knowledge and interaction history. Soc. Psychol. Q. 44, 42-48 (1981).

Polsky, R.H. and McGuire, M.T. Ethological analysis of manic-depressive disorder. J. Nerv. Ment. Dis. 167, 56-65 (1979).

Polsky, R.H. and McGuire, M.T. Naturalistic observations of pathological behavior in hospitalized psychiatric patients. J. Beh. Assess. 3, in press (1981).

Plutchik, R. *The Emotions*. Random House, New York (1962).

Plutchik, R. A general psychoevolutionary theory of emotion, in *Emotion: Theory, Research and Experience*, R. Plutchik, ed. Academic Press, New York (1980), pp. 3-33.

Stizer, M.L., Griff, R.R., Bigelow, G.E., and Leibson, I. Human social conversation: Effects of ethanol, secobarbital and chlorpromazine. *Pharm. Biochem. Beh.* 14, 353-360 (1981).

Trivers, R.L. The evolution of reciprocal altruism. *Q. Rev. Biol.* 46, 35-57 (1979).

Wilson, E.P. *Sociobiology: The New Synthesis*. Harvard University Press, Cambridge (1975).

SUBJECT INDEX

Acetylcholine, 316
Aggression,
  between males, 35, 36
  effects of Methaqualone, 218, 221, 223
  effects of THC, 76, 77
  effects on plasma testosterone, 41
  in dominant female talapoins, 43
  in dominant males, 41
  indirect effects of, 48
  in subordinate males, 41
  receipt of, 36, 43, 48
  territoriality, 7
Alcohol
  biphasic response, 265
  depressive effects of, 264
Amygdala
  affected by Methaqualone, 231
  lesions of, 2
  nucleus, in affiliative bonds, 4
Amphetamines
  agonistic behavior, 263
  altered maternal behavior, 241
  despair behavior, 263
  infant reaction, 241
  social isolation, 238, 241
  stereotypy, 238, 241
AMPT
  inhibitor of catecholamine synthesis, 271-273
  effects of social rank, 228
  effects on social interactions/behavior, 273
  reversible effects
Androgen level
  in castrated males, 16
Antidysphoric strategies, 289
  benzodiazepines, 289
  food, 290
  interpersonal relationships, 290
  other chemical agents, 289
  physical activities, 290

Autism, 244
  animal model, 244
Biopsychosocial balance, 281
Blood serotonin levels
  after separation from mother, 252
  in dominant males, 85
  influence on behavior, 90, 91, 96
  in subordinate males, 86
Bonding
  affectual, 284
  coitus as bonding function, 167
  grooming as social bonding, 128, 129
  relationships for reproduction, 170
  social, 128, 129
Catecholamine synthesis, 271-273
  inhibited by AMPT, 271-273
  specific neurotoxin (6 hydroxydopamine), 275
Chlorgyline
  MAO inhibitor, 90, 91
Chronotropism, 287
CNS depressants, 228
  causing behavioral changes, 228
  context-dependence, 307-309
Contraceptives, 11, 112, 135-139
  female control of reproduction, 136, 137
  impact on behavior of males, 136
  impact on female sexual behavior, 135
  impact on relations between females, 136
  loss of libido, 11
Copulatory behavior, 159, 160, 164, 169
  as bonding function, 167, 170
  as social action, 167, 170
  completed/uncompleted, 158, 159
  during menstrual cycle, 158
  during midcycle peak, 158, 159
  effects of kinship, 160, 162
  effects of rank, 160, 167
  female attractiveness, 160
  male-male competition, 160
  rates in MPA treated/untreated females, 114, 115, 125, 126
Cortisol, 12-15, 17, 19, 21-23, 65, 183, 252
  in subordinate animals, 24
  high/low responders, 60
  stress response, 59

Cost/debt, 333, 334, 336, 337, 338, 342
Cyproterone acetate, 175-179, 181, 183, 185, 199
  decreased testosterone level, 196-198
  effects on libido, 196
Depression, animal model, 249, 253
  altered sleep patterns, 252
  and alcohol, 264, 265
  and anxiety, 330
  and peer separation, 255
  and separation, 249
  basic causes, 330
Despair, 251, 252, 255, 257-260
  AMPT potentiation, 261
  following separation, 250, 251
Distance regulating mechanism, 297
Dominance, 9
  after estrogen treatment, 38-40
  and kinship, 47-51
  changing rank, 38-40
  correlation with size, 66
  defined by direction of aggression, 48
  determined by direction of aggression, 35, 36
  effect on endocrine physiology, 22-24
  effect of Methaqualone, 219
  effect of separation, 168
  effect on testosterone levels, 23, 25
  interacting with biorhythms, 25
  intermale, 65-67
  linear hierarchy, 10, 18, 20
  relations
    male-female, 147
    mother-son, 147
    brother-sister, 147
  sexual, 160
Dominance hierarchy, 10-14, 17, 35, 36, 254, 277
  in all-female group, 147, 168
Dominant male, 24
  blood serotonin in, 85
  changes caused by stress, 12
  reproductive potential, 50
  response to tryptophan, 95
  sexual invitations received, 45, 46
  status influence, 18
Dopamine
  depletion by AMPT, 271
  hypothesis of schizophrenia, 237

Drug effects
  in social contexts, 329
  in untreated animals, 327
  on benefit/cost calculations, 346
  on individuals, 346
  on systems, 346
Drugs
  Alcohol, 264, 265
  Alpha-Methyl-P-Tyrosine (AMPT), 228, 262, 271-273
  Amphetamine, 229, 237, 238, 241, 261, 263, 271-273
  Benzodiazepines, 289
  Chlorgyline, 90-91
  Cyproterone Acetate (CA), 175, 176, 178-199
  Elavil, 346
  Ethanol, 228
  Fluoxetine, 96-101
  Fusaric Acid (FA), 262
  Imipramine, 263-264
  Medroxyprogesterone Acetate (MPA), 3, 108-115, 117, 119, 121, 122, 125-126, 128
  Methaqualone, 3, 205-232
  Morphine, 228
  Naloxone, 276
  Parachlorophenylalanine (PCPA), 87-89, 91, 262, 310
  Pentobarbitol, 228
  Phenylzine, 287, 288, 304, 307, 310
  Pimozide, 243
  Progesterone, 107, 109-112, 117, 121
  Ritalin, 290
  Tetrahydrocannabinol (THC), 57, 71-74, 76-78, 80
  Tryptophan, 89, 95-102, 346
Dyadic interactions, 321, 324, 327, 329, 331, 334, 338-341, 343, 344, 347, 348
  feedback relationship, 327
  McKay-Kummer model, 339
Dysphoria, 285, 290, 293, 295, 302
  hysteroid, 299, 300, 312
Ejaculation rate
  altered by MPA, 119, 121, 124
Estrogen-treated females, 34, 35, 45
  rank changes in males, 38-40
Evolutionary theory, 339, 340

Family systems
  behavioral effects on, 291
  biopsychosocial balance, 281
  distance regulating mechanism, 297
  dysfunctions, 294
  patterns, 312
  receptors, 316
  resistance to change, 312
Fear provoking stimulus, 276
Gonadal function
  suppression of, 23
Grooming
  affected by MPA treatment, 119, 128
  as social bond, 128, 129
  duration of female to male, 119
Homeostatis, 283, 284, 291, 294, 298, 305, 311
5-HTP, 89-91
Hyperactivity, 290
Hyperkinetic syndrome, 291
  treatment, 290
Hypomania, 334
Hysteroid dysphoria, 299, 300, 312
Internal Feeling State (IFS), 281, 284, 285, 293
Medroxyprogesterone acetate, 3, 109, 111, 112
  effects of, 121, 125, 126
    on female attractiveness, 190, 126, 127
    on female presenting, 110
    on female proceptive behavior, 108, 110, 127
    on socio-sexual behavior, 109, 117, 127
  implants, 11
  influence on copulation, 114, 117, 126
  mean ejaculation rate, 119, 126
Menstrual cycle
  changes in social and sexual behavior, 107, 147, 154, 164
  copulatory behavior, 160, 163
  female attractiveness, 143, 157, 165, 167
  grooming of males, 147
  kinship, 147, 157
  midcycle peak, 111, 158, 159
  proceptivity, 108, 110, 127, 143
  sexual activity, 108, 141, 158
  receptivity, 144, 165
Methaqualone, 3, 205-232
  aphrodisiac effect, 205, 207, 232
  hypnotic effect, 205
  mood, 281, 284-286
  social rank, 226

Neuroleptics, 237, 244
  effects on social behavior, 242
Norepinephrine, 262, 271
Parachlorophenylalanine, 310
  activity, 89
  effects of, 87, 89, 91
  properties, 88, 89
Peer relations
  lifetime, 254
  peer separation, 255, 257, 259, 260
Power-dependency theory, 339, 340
Progesterone, 107, 109, 110, 117, 121
  contraceptives, 111, 112
  decreased female proceptivity, 110
  loss of libido, 111
Protest stages, 250-252, 255, 257, 258
Proximity, 204, 205, 281, 284-286, 295, 302
Rank
  aggression in, 36, 41
  AMPT effects
  change after testosterone, 36
  change in male with estrogen-treated female, 40
  copulatory behavior, 160, 167
  effects of Methaqualone, 226
  reproductive advantage of, 169
Seasonality, 13-17
Separation
  effects of age and social conditions, 251
  effects of drugs, 262
  effects of neurochemical systems, 252
  effects of separation
    biphasic reaction, 251
    despair, 250, 265
    protest, 250, 255
    withdrawal, 251
  effect of sex and preexisting behavior patterns, 251
  effects on sleep patterns, 252
  grief reaction, 254
  mother-infant separation, 250-252, 265
  neurobiological variables, 261, 265
  reactions of older animals, 258
  severity of response, 256
  sex of infants, 251, 252
  weaning, 59

Serotonin
  see blood serotonin
Stereotypy, 237, 238, 241
Stress
  enhancement by THC, 76
  neuroendocrine, 68-70, 79
  post-stress testosterone levels, 16
  psychosocial, 57
  separation, 59
  testosterone suppression, 67
Subordinate male, 8, 12, 13
  aggression, 40, 41, 43
  blood serotonin, 86
  influence of social constraints, 51
  reproductive potential, 50
  testosterone levels, 45
Temporal pole lesions, 2
Testosterone levels, 12-15, 19, 21-23, 25, 33, 34, 36, 38, 40, 45-57, 178, 181, 183, 196
  effects of aggression, 41, 50
  effects of attractive female, 38
  effects of housing conditions, 38
  in subordinate males, 45
  post-stress, 16
  rank-related differences, 36
  sexual activity, 49
  spermatogenesis, 50
  suppression, 16
Tetrahydrocannabinol
  CNS tolerance, 71-73, 76
  effects on
    adaptive behaviors, 71
    aggression-enhancement, 76, 77
    inter-individual variability, 71
    long-term exposure, 74, 77
    reversal of effects, 71
    social behavior, 57, 71, 72, 80
    social interactions, 76
  stress-enhancement, 76
  tranquilizing effects, 76, 77
Tryptophan, 346
  behavioral effects, 96-101
  effects, 102
  primary sites of action, 89